100 Ready-to-use Pathfinders for the Web

A GUIDEBOOK AND CD-ROM

A. PAULA WILSON

NEAL-SCHUMAN PUBLISHERS, INC.
NEW YORK LONDON

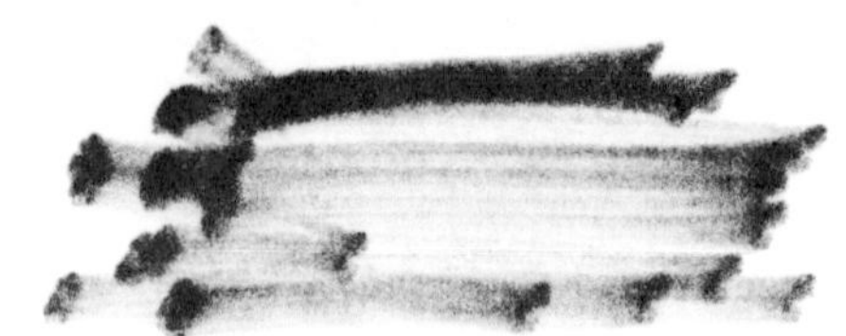

Published by Neal-Schuman Publishers, Inc.
100 William Street, Suite 2004
New York, NY 10038

Printed and bound in the United States of America

The paper used in this publication meets the minimum requirements of American National Standard for Information Sciences—Permanence of Paper for Printed Library Materials. ANSI Z39.48-1992. ∞

Library of Congress Cataloging-in-Publication Data

Wilson, A. Paula.

100 ready-to-use pathfinders for the Web: a guidebook and CD-ROM / A. Paula Wilson.
p. cm.

Includes bibliographical references and index.

ISBN 1-55570-490-5 (alk. paper)

1. Internet research—Handbooks, manuals, etc. 2. World Wide Web—Subject access—Handbooks, manuals, etc. 3. Reference services (Libraries)—Handbooks, manuals, etc. 4. Bibliography—Methodology—Handbooks, manuals, etc. I. Title: One hundred ready-to-use pathfinders for the Web. II. Title.

ZA4228.W55 2005
001.4'2'02854678—dc22 2004040292

For my Mom and Dad,
Mr. Louis E. and Marie Azar of Great Road

Table of Contents

Part II: 100 Ready-To-Use Pathfinders

List of Figures

Preface

Imagine having ready-to-go answers to your most frequently asked questions—consumer affairs, quotations, government information, diseases, genealogy—and for those questions that stop you in your tracks—American folklife, military science, decorative arts identification, small business start-up, biomes, saints. Better yet, imagine a library tool that can reach your users before they reach the reference desk. *100 Ready-To-Use Pathfinders for the Web: A Guidebook and CD-ROM* helps your library utilize valuable research aids that anticipate information needs, identify resources for specific subjects, link directly to information, and promote the library's holdings, all from your library Web page. It provides you with advice for designing, building, implementing, and collecting these time-saving tools on your own.

The book is divided into two complementary parts. Part I, "Creating Pathfinders," outlines the use, construction, and purpose of these finders in today's modern information setting. Chapter 1, "Exploring Pathfinders," provides definitions and history; outlines the use and users of the Web-based version; offers tips for increasing accessibility and utilizing promotion; and describes some of the cooperative efforts and commercial ventures that are making these tools more valuable to today's libraries. Chapters 2, "Writing Pathfinders," and 3, "Presenting Pathfinders on the Web," offer suggestions for creating and formatting your own findings aids, including selecting topics, gathering content, organizing resources, increasing usability, hypertext linking, and maintaining your collection.

Part II, "100 Ready-To-Use Pathfinders" presents one hundred subject guides. The selected topics cover some of the most common and specialized searches requested for not only educational research, but also practical, everyday inquiries like elections, travel, and job-hunting skills. They contain some of the best sources for information on their specific subjects and allow modification of their contents to reflect the services, collection, and population of your library. You can incorporate local content, related program information, special collections or unique materials, links to local organizations and Web sites, or licensed databases. These premade guides are a perfect way to start or supplement a collection of pathfinders.

100 Ready-To-Use Pathfinders for the Web includes a CD-ROM with all of the pathfinders reproduced in both Microsoft Word and XHTML formats with cascading style sheets for easy upload to library Web sites. The style sheet can be modified to reflect specific color choices or font specifications by using Web-editing software. Step-by-step instructions for use appear on the CD-ROM.

Because the most valuable pathfinders include those that list local information, the accompanying CD-ROM includes easy-to-follow instructions for creating new pathfinders, improving existing ones, or customizing the pathfinders available in the CD-ROM. The intent of this book is to offer librarians tools to jump-start their pathfinder collections. Librarians may chose to post the pathfinders in their current format, or begin customizing them one by one. Pathfinders were compiled by the author, research assistants, and/or thirteen contributing libraries and have been supplemented with further resources and annotations. Pathfinders chosen for the project were exemplary in content and/or format, and appear in screen shots throughout the book or, in some cases, have been adapted, condensed, revised, or augmented to provide the pathfinders that appear in Part II of the book.

Many challenges arose during the creation of this book. Most obvious are those associated with publishing (in print) information intended for online publication. For example, how do you best display a Web site if the link is embedded in the text? Another challenge arose when deciding which print materials to include, or which subscription databases to exclude, and how to treat materials that may not be common to all collections. Each challenge was solved, and maybe not entirely, by keeping in mind the intent of the book and how the reader would benefit most. Although this manual aims to assist librarians in the creation of pathfinders, the author acknowledges that time and effort is still needed to make these pathfinders their most useful by including resources specific to the library's geographic location or its local collections.

Creating effective, well-organized, and easily accessible pathfinders is an excellent way to relieve some of the stress on the reference desk, while also increasing your users' satisfaction with the library. The library Web site should be a functional tool for researchers—complete with not only the catalog, databases, and indexes, but also materials that aid in research and allow patrons to work independently. These guides are the perfect way to get your patrons looking in the right direction and to make them more aware of the full resources that your institution offers. *100 Ready-To-Use Pathfinders for the Web: A Guidebook and CD-ROM* is a resource for all libraries interested in better serving their patrons and better managing their professionals' time. I hope that these pathfinders find their way onto your Web site and that they inspire you to create all new resources to meet the needs of your community.

Acknowledgments

I would like to thank my editor Michael Kelley whose encouragement and guidance proved invaluable as the project offered challenges along the way. At each turn in the development of this book, we faced challenges, and Michael was there to help me rough out the edges. Corrina Moss, Publishing Assistant, and Miguel Figueroa, Assistant Director of Publishing, provided the momentum that kept this book moving through the production process. Additionally, I would like to thank Charles Harmon, Director of Publishing at Neal-Schuman, who "saw" this book long before I did. His insight early on provided me with the roadmap I needed to develop the book's concept.

This book would not exist without the expert assistance of librarians Yvette Dowling and Cindy DeLanty, who assisted in identifying libraries that had outstanding pathfinders in specific subject areas. They also spent time editing and revising pathfinders. Their commitment to the project, especially as it neared completion, is greatly appreciated. I would also like to acknowledge former colleagues from the Las Vegas-Clark County (NV) Library District. Especially helpful were Lisa Spurgin, Isabel Boylan, Gregory Robinson, Vickie Barnett, Pam Zehnder, Jan Passo, and Nancy Ledeboer. Their ideas and conversation helped develop the InfoGuides at that library which ultimately lead to the creation of this book.

Because the most helpful pathfinders are best written by librarians with a subject emphasis or passion for a particular topic, other libraries were invited to contribute to the book. Some materials in this book have been adapted from fifty-five pathfinders collected and used by generous permission from the following libraries:

Arthur Lakes Library, Colorado School of Mines
www.mines.edu/library/reference/subject.html
General Engineering, Chemistry and Chemical Engineering,Geology

Fairfax County (VA) Public Library
www.co.fairfax.va.us/library/Homework/Research/default.htm
Ancient Cultures, College Search, Mythology, World War I, World War II

Indianapolis-Marion County (IN) Public Library
infozone.imcpl.org/kids_pathfinders_alpha.htm
Elections, Exploration and Discovery, Space Shuttle, Summer Olympics, Winter Olympics

Internet Public Library
www.ipl.org
Antiques and Collectibles Appraisal, Consumer Information & Advocacy, Etiquette, Genealogy, U.S. Income Tax Preparation

Las Vegas-Clark County (NV) Library
www.lvccld.org/ref_info/info_guides/index.htm
Finding Government Information, African American history, Asian Pacific American heritage, National Hispanic Heritage, Native American Heritage

Los Angeles (CA) Public Library
www.lapl.org/inet/index.html
Automobile Repair Information, Standards and Specifications

Massillon Public (OH) Library
www.massillon.lib.oh.us/ya/homework/pathfinders.htm
Presidents, Decades, U.S. Civil War, Biomes, Ancient Egypt

Morris Library, Southern Illinois University, Carbondale
www.lib.siu.edu/cgi-bin/encore2/resources
History, Demographics, Business, Management, Marketing Resources

New York Public Library
www.nypl.org/research/electronic/subject.cfm
Biographical Resources, Book Reviews, Identifying Objects, Obituaries, Death Notices, Guides to Burial Places of Notables, Quotations

Olin Library, Cornell University
www.library.cornell.edu/olinuris/ref/subguides.html
Anthropology, English literature, History, Multicultural Literature in the United States, Psychology

Ralph Brown Draughon Library, Auburn University
www.lib.auburn.edu/scitech/resguide/computer/compusci.html
Computer Science and Software Engineering

University Libraries, University at Albany, SUNY
http://library.albany.edu/subject/
Criminal Justice, Education, Public Administration and Policy, Social Welfare, Theatre

Wilson Library, University of Minnesota
www.lib.umn.edu
Materials adapted from LUMINA®, the University of Minnesota Libraries Web site.
Geography, Law, Linguistics, Military Science, Women's Studies

Last, but certainly not least, I would like to thank my husband and "Director of Common Sense," Brian Wilson, who became my pathfinder during this writing project, guiding me each step of the way in less than two pages.

Anatomy of a Pathfinder: Bibliographic Units and Other Categories

A list of bibliographic categories can assist librarians in the preparation of creating pathfinders. Inclusion of each category is dependent on the topic and type of library so all units listed below may not be used in one pathfinder. The list below is not intended to represent how the items will appear on a Web page, but to provide a librarian with guidance for gathering resources. See the two sample pathfinder templates on pages xxii–xxiii for suggested placement of each category.

1. Library's Logo

Place the library's logo in the upper left corner of the page. The logo should link back to the library's homepage.

2. Library's Main Navigation Bar

Place the library's main navigation bar on the pathfinder so that visitors can link back to other parts of the library's Web site. Web users should have a clear notion of where they are in the site. The navigation bar will help them to move back to the library catalog or to other parts of the Web site. Main navigation bars are typically placed on top of the page or on the left-hand side.

3. Title: A Research Guide

Place the topic title prominently at the top of the page. A subtitle may be added to provide further description of the document's function.

4. Table of Contents

A table of contents or navigation bar should be placed on the top of the page so that users can link from each bibliographic category to its location further down on the page.

5. Introduction

In about 3–4 sentences, provide a brief explanation of the topic and specific highlights as they relate to the topic.

6. Print a Printer-Friendly Page

Libraries that use many graphics on their pathfinders may want to offer the option of printing a graphics-free pathfinder. A graphical button or text indicates the availability of a printer-friendly page.

7. Local Information

The most valuable pathfinders are the ones that include local information. Libraries can segment this type of information to one area (as done in this book) or include it within other categories.

Programs and Events

List library events such as tax preparation services or genealogy classes when relevant to a topic.

Local Agencies and Organizations

List contact information and descriptions of local agencies that may be of interest to users of this pathfinder.

Local Resources

Post local services or publications related to the topic.

Course Pages or Academic Departments

List course pages that relate to the topic of the pathfinder. In many cases teachers or professors have constructed accompanying Web pages for their classes. Additionally, academic departments usually have a Web page detailing their program and classes.

Related Pathfinders

Use this opportunity to market other pathfinders that may be useful to your visitors.

8. Research Materials (Books, videos, DVDs, CDs, etc.)

Create links to circulating materials in the library catalog. Customers may find it useful if you provide links to various formats of books such as audio books, documentary films, or music selections. Include cover art if possible. Hypertext links can take your users directly to the materials in the catalog, where they can place holds or requests on the items.

Encyclopedias and Dictionaries

List general and subject-specific materials that provide an overview of the topic.

Handbooks and Manuals

Include any related materials that instruct or guide the researcher.

Directories and Sourcebooks

Include subscription and free Web-based directories (such as *Reference USA* and local chamber of commerce business directories) or books that provide referral information for organizations, products, and companies.

Electronic Journals

Libraries may subscribe to individual electronic journals or may have access to them through aggregated subscription databases. A link to the journal title normally displays a table of contents users can select a particular date or search by keyword, title, author, and subject.

Magazines and Newspapers

Some of the major vendors allow deep linking to full-text articles or to journal and magazine table of contents. The vendor can work with libraries to solve access problems related to patron authentication.

Online Databases

List links to relevant databases with descriptions that are specific to each topic. For example, in your employment pathfinders, explain how visitors can use a business directory to locate potential employers.

E-books

Use cover art and annotations to display related e-books.

Web Sites

Select Web sites that meet collection development criteria. Create embedded links within the title of the Web site. Some libraries may choose to display the full Web site address; however, lengthy addresses can be cumbersome to display.

Associations

List associations, their contact information and links to their Web sites.

Primary Sources

Information that provides first-hand account or original data such as journal entries, newspaper articles, interviews, and government document.

Biographies

List any books detailing the life of an individual or list collective biographies such as *Asian-American Authors*, by Kathy Ishizuka.

Statistics

List here books or Web sites that offer statistics on the subject. For example, if the topic is the library's home state of Texas, a link to State and County QuickFacts at the U.S. Department of Census site brings researchers directly to a page that offers statistics about Texas (http://quickfacts.census.gov/qfd/states/48000.html).

Preprints

A preprint is anything that appears before the official printing-often a scientific article that will later appear in a journal. Preprints are normally peer reviewed and scientific and technical in nature.

Indexes and Abstracts

An index is a finding aid that provides access to a field's publications arranged by author, subject, title, and other categories. An example is *Library Literature, Biography Index.* Abstracts—objective summaries of publications—may be included.

Bibliographies

Bibliographies offer researchers suggestions for further materials once they have exhausted those available on the pathfinder. Bibliographies can be found in the library catalog, other catalogs, or even on the Web.

Citation Styles

Libraries can include a brief listing of the more popular citation styles or more detailed information on the one most preferred by the affiliated school. Here is an example of how this information can be displayed:

The Chicago Manual of Style. 15th ed. Chicago: University of Chicago Press, 2003.
An essential reference for authors, editors, proofreaders, indexers, copywriters, designers, and publishers in any field, it covers publishing formats, editorial style and method, documentation of electronic sources, and book design and production.

A Manual for Writers of Term Papers, Theses, and Dissertations, 6th ed. by Kate L. Turabian. Revised by John Grossman and Alice Bennett. Chicago: University of Chicago Press, 1996.
B. Honigsblum's revised and expanded version of Kate Turabian's standard guide for student writers. Fourteen chapters cover how to put a paper together, from introductory chapters to the bibliography.

MLA Handbook for Writers of Research Papers, 6th ed. by Joseph Gibaldi. New York: Modern Language Association, 2003.
The *MLA Handbook* covers all aspects of research writing, from selecting a topic to submitting the completed paper. Aimed at high school and undergraduate students.

MLA Style Manual and Guide to Scholarly Publishing, 2nd ed. by Joseph Gibaldi. New York: Modern Language Association, 1998.
The standard guide for graduate students, teachers, and scholars. Includes citation and stylistic conventions in the preparation of manuscripts, theses, and dissertations.

Publication Manual of the American Psychological Association, 5th ed. Washington, DC: American Psychological Association, 2001.
Provides guidance on grammar and a reference and citation system and how to accurately publish numbers, metrication, statistical and mathematical data, tables, and figures for use in writing, reports, or presentations.

Frequently Asked Questions

Some pathfinders, in addition to listing resources, may list frequently asked questions about certain topics. For example, some of the frequently asked questions of a patron starting a business may include the following: How do I write a business plan? Where can I get money for my business? How do I register my business?

Subject Headings and Keywords

Libraries may display and link to Library of Congress Subject Headings, Dewey classifications, or Sears List of Subject Headings. Insert appropriate subject headings and keywords that, when clicked, generate a search in the library catalog. These links are especially useful because they produce dynamic results as material is continually added to the library catalog. Explore the functionality of the library's Web catalog. For example, some library catalogs allow current awareness or selective dissemination of information (SDI) services so that customers can run an automated search at a frequency they specify.

Call Numbers to Browse

Offering call number ranges can help direct customers to the right section of the library. Depending on the topic, inclusion of call numbers may either help or hinder in locating the items. For example, libraries within a multi-branch system may have different locations for the item, making it difficult to locate without viewing the detailed holdings.

9. Copyright Notice and Date Created or Revised

The copyright notice gives credibility and states ownership. The date will show users its currency.

10. URL for the Pathfinder

Including the URL of the pathfinders is important so that when a visitor prints out the pathfinder, the Web address will appear on the printed copy.

Adapted from Wilson, Paula. 2002. "Perfecting Pathfinders for the Web." *Public Libraries* 41, no. 2 (March/April): 99–100.

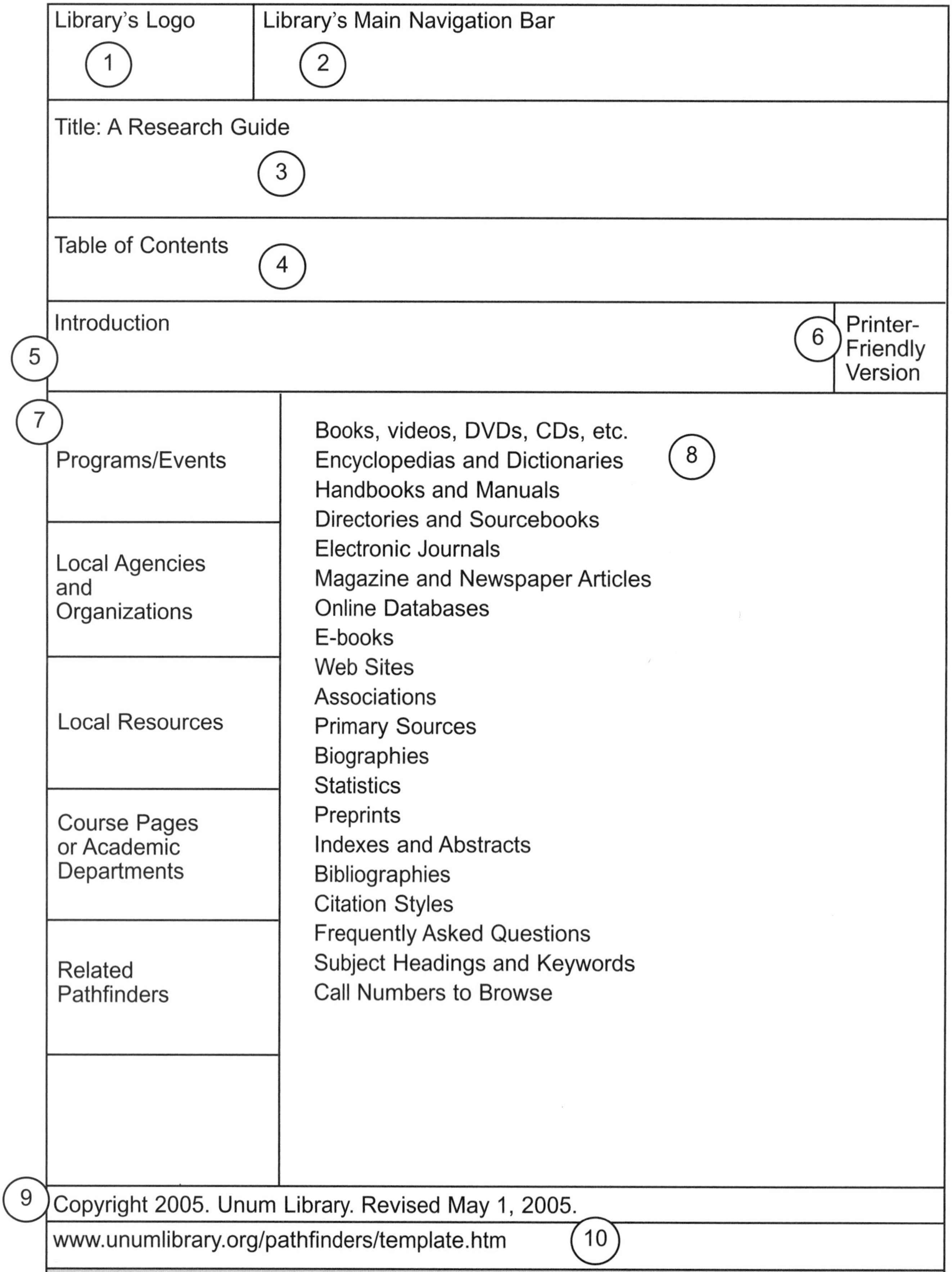

Library's Logo (1)	Library's Main Navigation Bar (2)
Title: A Research Guide (3)	
Table of Contents (4)	
(5) Introduction	(6) Printer-Friendly Version
(7) Programs/Events	(8) Books, videos, DVDs, CDs, etc. Encyclopedias and Dictionaries Handbooks and Manuals Directories and Sourcebooks Electronic Journals Magazine and Newspaper Articles Online Databases E-books Web Sites Associations Primary Sources Biographies Statistics Preprints Indexes and Abstracts Bibliographies Citation Styles Frequently Asked Questions Subject Headings and Keywords Call Numbers to Browse
Local Agencies and Organizations	
Local Resources	
Course Pages or Academic Departments	
Related Pathfinders	
(9) Copyright 2005. Unum Library. Revised May 1, 2005.	
www.unumlibrary.org/pathfinders/template.htm (10)	

Template 1: Anatomy of a Pathfinder

Library's Logo (1)	Library's Main Navigation Bar (2)
Title: A Research Guide (3)	
Introduction (5)	
Table of Contents (4)	Printer-Friendly Version (6)

Books, videos, DVDs, CDs, etc.
Encyclopedias and Dictionaries
Handbooks and Manuals
Directories and Sourcebooks (8)
Electronic Journals
Magazine and Newspaper Articles
Online Databases
E-books
Web Sites
Associations
Primary Sources
Biographies
Statistics
Preprints
Indexes and Abstracts
Bibliographies
Citation Styles
Frequently Asked Questions
Subject Headings and Keywords
Call Numbers to Browse

Programs/Events
Local Agencies and Organizations
Local Resources (7)
Course Pages or Academic Departments
Related Pathfinders

Copyright 2005. Unum Library. Last updated: May 1, 2005. (9)

www.unumlibrary.org/pathfinders/template.htm (10)

Template 2: Anatomy of a Pathfinder

Part I
Creating Pathfinders

1
Exploring Pathfinders

Definition

Pathfinders organize annotated lists of topical resources from a variety of formats and sources in one location. Pathfinders are suitable for beginning researchers and are not intended to be exhaustive, but rather a starting point for the beginning researcher within a topic. As defined by Dunsmore, "Pathfinders arrange information resources in a search strategy order, in order to introduce a researcher to an unfamiliar topic and to facilitate access to information. Pathfinders are bibliographic guides that list selective resources of any format, provide location information (i.e., call numbers and library locations), and annotate these resources to differentiate them for the user. Synonymous with subject guides" (Dunsmore, 2002: 150). Dunsmore's research focused on Web-mounted, academic business pathfinders and served to bridge a gap in the literature due to the transition of paper pathfinders to electronic ones. Additionally, at a time when many print materials were transformed to Web-formatted resources, the ability to provide access to them through hypertext links greatly improved the usefulness of library pathfinders.

Pathfinders facilitate research by instructing, informing, and promoting awareness of related resources about which information seekers may otherwise not be aware. Additionally, pathfinders designed for the Web provide guided searching by incorporating hypertext links that connect researchers to resources or even execute a predefined search in a remote database. Other related terms include research or subject guides, bibliographies, and webliographies.

A Brief History

Printed guides, in the form of reading lists, proved useful when patrons were first allowed access to what were once closed stacks. Additionally, librarians relied even more on reading lists as collections grew and library patronage increased due to an emergent literate population (Dunsmore, 2002). These guides offer a way for librarians to assist customers in order to highlight collections or provide instructional material to guide users through a variety of frequently requested research queries. Additionally, school libraries prepare bibliographies for class visits or professors. Dunsmore points out that Samuel Swett Green recognized and advocated the library's role of selecting and recommending books for readers (Dunsmore, 2002: 138).

Furthermore, Dunsmore's research concludes that reading lists, book lists, or printed guides appear to be predecessors to pathfinders, although the term does not appear in the library literature until 1972.

Pathfinders: Who Benefits from their Use

Although, at least initially, it may take a good many staff hours to create and format pathfinders, their impact level is high and immediate, especially when they are made available on the Web, because Web technology offers so many new possibilities for libraries to connect researchers with resources. Students and researchers, teachers and faculty, and librarians all use pathfinders. These tools meet the needs of remote users and distance education students and printer-friendly versions of pathfinders can also be created to meet the needs of researchers in the library.

Students and Researchers

Students and researchers use pathfinders as a tool to acquaint themselves with an unfamiliar topic. Patrons expect pathfinders to point them to the standard sources covering a topic, to distinguish one resource from the other by reading annotations of resources, so they can easily locate the material they seek. Researchers typically prefer to work with online full-text information; however, it is equally important to offer researchers the most relevant sources regardless of format.

Teachers and Faculty

Educators—schoolteachers and college and university professors—may work in conjunction with the school or university librarian to produce pathfinders, which facilitate curriculum-based research. By including a section about evaluating sources and by highlighting tools on how to find more information, the pathfinder can be used as a tool for information literacy. Additionally, these pathfinders would ideally include information on preferable citation styles. In some instances, course pages are created to facilitate research for a particular class and include links to related academic departments. Pathfinders in schools are designed not only to connect researchers with the material they need, but also to educate them in the research process, the evaluation of information, and how to present the resources used through bibliographic citations.

Librarians

Librarians use pathfinders for a variety of reasons although they are created principally for their customers. Librarians use them to provide both individuals and groups with a tool to explore the collection for frequently requested topics or special collections. They are also designed as a time-saving tool since pathfinders represent the most frequently requested questions thereby avoiding duplication of effort. For example, the steady requests for genealogical information typically found at many public libraries can be filled quite easily by a pathfinder designed specifically for the budding genealogist. Pathfinders are also useful for staff unfamiliar with special topics, such as law, medical, or business information.

Librarians may use pathfinders as a collection development tool or as a training tool to introduce new librarians to local resources on popular topics. Additionally, library systems with central and branch libraries benefit from publishing online pathfinders to promote an awareness of materials not easily located in the catalog or not cataloged, and materials purchased only for a central branch or special collections.

Uses of a Pathfinder

Pathfinders are created for a variety of reasons such as class visits to the library, perhaps at the request of a faculty member, or to avoid redundancy and duplication of efforts by staff when answering frequently asked questions. Pathfinders also provide access to library materials that may otherwise be unknown, as well as a way for librarians to provide reference and instruction services to patrons who may never even come into the library.

Awareness and Promotion of Library Collections

Many times, topical materials are not shelved in the same physical location sometimes because of an item's status or format. For example, library materials such as maps, government documents, and multimedia, or items in special collections may be located in different areas of the library. Pathfinders provide a tool for librarians to bring all of the formats together in one resource list. Pathfinders can be used to promote print and online resources about which library users may not be aware, in addition to external resources such as local or national organizations.

Reference Services

Many reference librarians enjoy creating pathfinders, because it is a large part of what they are trained to do: locate relevant material, compile resources, and write annotations.

There are at least five items listed in the Reference and User Services Association's (RUSA) *Guidelines for Information Services* that support the creation of pathfinders:

1. Make available user aids to identify items in the collection relevant to their interests and needs.
2. Provide instruction in the effective use of its resources by, for example, creating guides to the collection.
3. Publicize the scope, nature, and availability of the information services it offers.
4. Collect and create information and referral files to provide access to the services and resources of local, regional, and state organization.
5. Provide information even if it has not been explicitly requested.

(Reference and User Services Association, 2001)

Libraries that maintain pathfinders meet many of the RUSA service goals listed above. Pathfinders can also facilitate reference transactions, especially online transactions. For example, if a librarian is helping a patron using online chat software and must place a customer "on hold," features in the software allow them to send hypertext links or push Web pages to the customer's desktop. If the patron is looking for biographical information, the librarian can send them the link to the pathfinder on biographies, while the librarian either finishes up with another customer or checks some other references. Additionally, patrons calling in by telephone whose question or topic can be answered or further explored by a Web-based pathfinder may be best served by providing them with the Web site to access the page.

Information referral files, once stored in a folder or file box, can be shared among staff and users. By incorporating this information into the appropriate pathfinder the librarian reduces the number of places that the researcher has to check.

Pathfinders may also fill the need of a user who may not even reach the reference desk. For example, since pathfinders are created to fulfill the information needs frequently requested by their customers, many questions are never initiated by the patrons. Their information need is met when they pick up a printed copy of the pathfinder from a display rack in the library's lobby, or by browsing the library's Web site.

With an increase in the number of students completing courses through distance education, pathfinders become even more valuable, especially if they focus on full-text online resources. Fortunately, many library materials have migrated from print to digital format making it easier for librarians to provide researchers with seamless access to many full-text materials. This anticipated yet unexpressed need, as the RUSA guidelines mention, can be fulfilled adequately by creating pathfinders.

Instructional Tool

Pathfinders offer a tremendous amount of potential as an information literacy tool, especially in school and academic libraries whose mission includes a large component of bibliographic instruction. Their purpose remains not only to provide the researcher with resources, but also to introduce students to the type of resources available in the subject area, as well as the research process itself.

Pathfinders also facilitate class visits or library assignments. When librarians collaborate with teachers they ensure that the target audience is well served, because the content is responsive to their needs both in the subject area as well as in the methodology of the research process. To make more efficient use of class time it benefits the students to review the pathfinder before the class visit. Because class periods are finite, pathfinders allow librarians to make productive use of time when the amount of material to be covered exceeds the time available (Graves, 1998). Pathfinders can also facilitate the research for distance education students and online classes who may never come into the library or work face-to-face with a librarian.

Merely offering links to resources does not offer the necessary skills researchers need to become at least comfortable, or more optimistically, proficient, in research. Pathfinders can increase information literacy skills by including the following categories: (1) evaluations of Web sites; (2) links to style manuals and writing guides; (3) links to various bibliographic citations; and/or (4) instruction on how to create Boolean searches, in addition to constructing them for the student through embedded searches within hypertext links (see Figure 1-1).

In a public library setting informational literacy skills may be more appropriately incorporated into research guides, which cover topics that involve in-depth and ongoing research, such as genealogy or patent searching.

Accessibility

Web users access information differently, and depending on their preference, may choose to browse or search for information. Accessibility of library pathfinders depends on including a label or terminology that refers to the entire collection, which researchers may find on the library's homepage or by searching the site. Herein lies the key to access. The term must be one recognizable to the researcher. For example, Dunsmore's research on business pathfinders indicates that although there was considerable variation in the terminology libraries used, all of the synonyms centered on the concepts of resources, research, subject, and guides. The most frequently used phrase in Dunsmore's research was "Research Guides." The

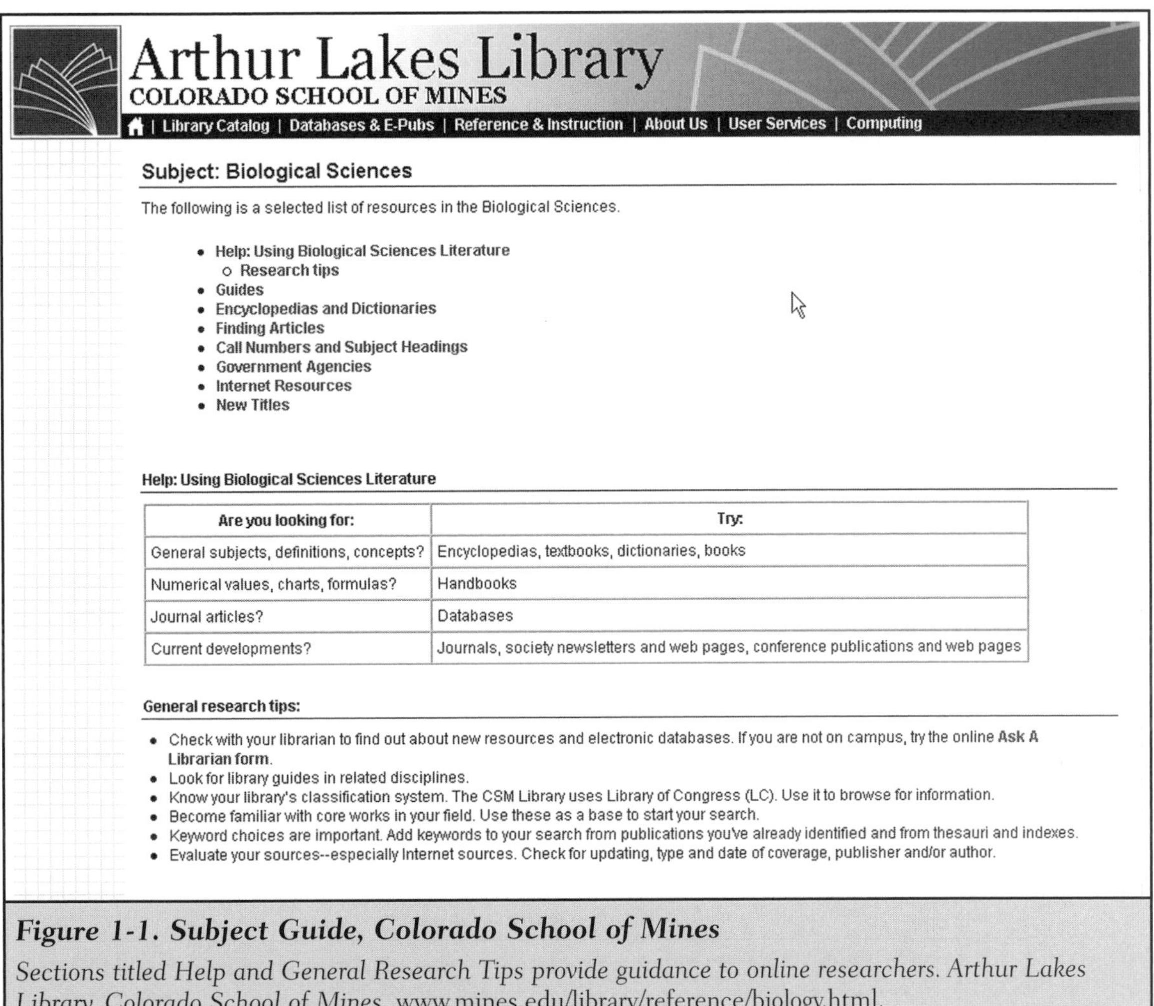

Arthur Lakes Library
COLORADO SCHOOL OF MINES

| Library Catalog | Databases & E-Pubs | Reference & Instruction | About Us | User Services | Computing

Subject: Biological Sciences

The following is a selected list of resources in the Biological Sciences.

- Help: Using Biological Sciences Literature
  - Research tips
- Guides
- Encyclopedias and Dictionaries
- Finding Articles
- Call Numbers and Subject Headings
- Government Agencies
- Internet Resources
- New Titles

Help: Using Biological Sciences Literature

Are you looking for:	Try:
General subjects, definitions, concepts?	Encyclopedias, textbooks, dictionaries, books
Numerical values, charts, formulas?	Handbooks
Journal articles?	Databases
Current developments?	Journals, society newsletters and web pages, conference publications and web pages

General research tips:

- Check with your librarian to find out about new resources and electronic databases. If you are not on campus, try the online **Ask A Librarian form**.
- Look for library guides in related disciplines.
- Know your library's classification system. The CSM Library uses Library of Congress (LC). Use it to browse for information.
- Become familiar with core works in your field. Use these as a base to start your search.
- Keyword choices are important. Add keywords to your search from publications you've already identified and from thesauri and indexes.
- Evaluate your sources--especially Internet sources. Check for updating, type and date of coverage, publisher and/or author.

Figure 1-1. Subject Guide, Colorado School of Mines

Sections titled Help and General Research Tips provide guidance to online researchers. Arthur Lakes Library, Colorado School of Mines, www.mines.edu/library/reference/biology.html.

term "pathfinder" has, for the most part, been used exclusively within librarianship and thus, is not familiar to the general public or students (Dunsmore, 2002).

Placement on the Library Web Site

Dunsmore's study (2002) of academic business pathfinders revealed that, most commonly, the library homepage was the level that first mentioned the link to access the pathfinders, but to arrive at the destination pathfinder required the user to keep clicking links to level four. Additionally, academic libraries or libraries with special departments and collections may house their collection of pathfinders on different sections of the Web site, which may further impede their access, especially if the authoring librarians use different terminology to label them. Libraries should use a recognizable label such as "research guides" and ensure they are available on or off of the main menu. That way, if a researcher enters the site from any other page than the homepage, she or he will still be able to find the pathfinders created to meet their information needs. Links to pathfinders should be placed on the research page, nearby or listed with licensed databases, and as new pathfinders are created, on the "what's new" page.

An Index Page

Once researchers click on the "Research Guides" label, they normally find an index page or table of contents that provides access to the collection of available pathfinders. In many cases, pathfinders are accessible by subject, alphabetically by title, and when the amount of pathfinders warrant it, a search engine (see Figure 1-2).

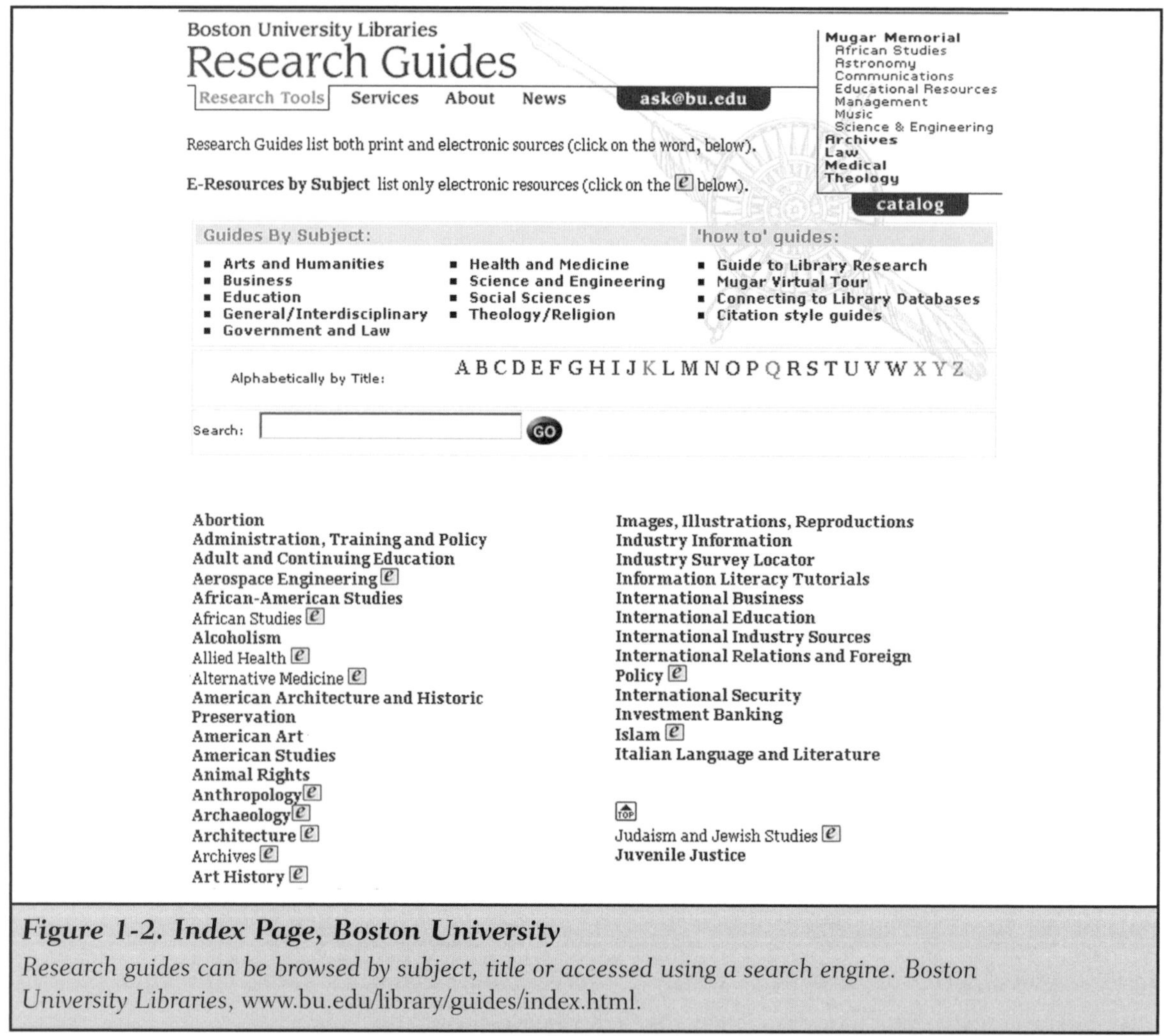

Figure 1-2. Index Page, Boston University

Research guides can be browsed by subject, title or accessed using a search engine. Boston University Libraries, www.bu.edu/library/guides/index.html.

Access through the Library Catalog

Pathfinders can also be cataloged in order to assist researchers who may do much of their research through the library's catalog. The 856 field of the MARC record contains the URL of the pathfinder's Web page. This way, pathfinders are retrieved from a subject search as illustrated in Figure 1-3. For example, a student working on a position paper about abortion searching the library catalog late at night may retrieve the library pathfinder if it has been cataloged. From there the student can continue linking through to relevant informational sources.

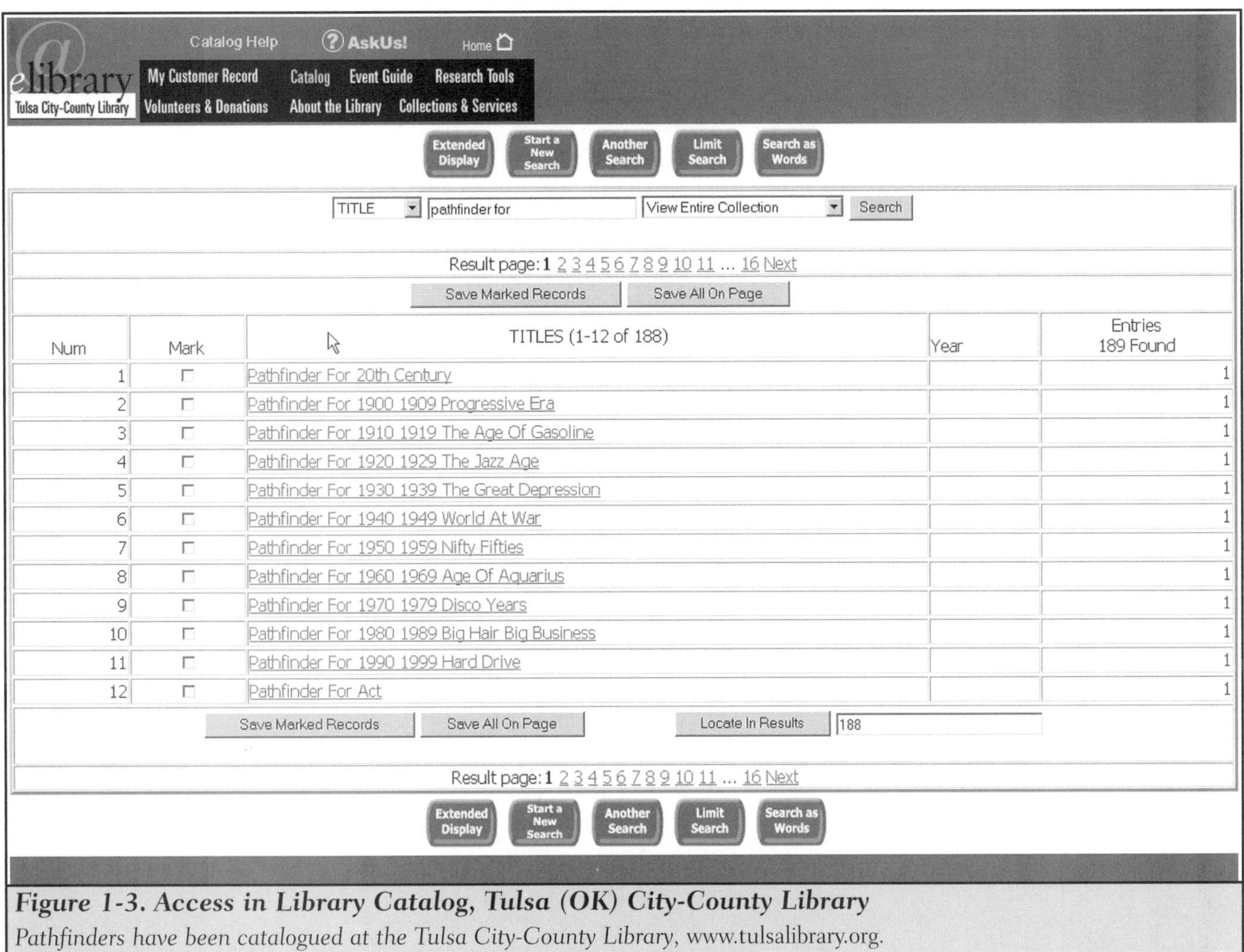

Num	Mark	TITLES (1-12 of 188)	Year	Entries 189 Found
1		Pathfinder For 20th Century		1
2		Pathfinder For 1900 1909 Progressive Era		1
3		Pathfinder For 1910 1919 The Age Of Gasoline		1
4		Pathfinder For 1920 1929 The Jazz Age		1
5		Pathfinder For 1930 1939 The Great Depression		1
6		Pathfinder For 1940 1949 World At War		1
7		Pathfinder For 1950 1959 Nifty Fifties		1
8		Pathfinder For 1960 1969 Age Of Aquarius		1
9		Pathfinder For 1970 1979 Disco Years		1
10		Pathfinder For 1980 1989 Big Hair Big Business		1
11		Pathfinder For 1990 1999 Hard Drive		1
12		Pathfinder For Act		1

Figure 1-3. Access in Library Catalog, Tulsa (OK) City-County Library
Pathfinders have been catalogued at the Tulsa City-County Library, www.tulsalibrary.org.

Promotion

How do researchers find library pathfinders? They can certainly show up as one of many pages returned from a search in any of the popular search engines, but how can the library make their own customers aware of the availability and usefulness of pathfinders? Informational needs are best met in context. The catalog is used as an access and promotional tool. Libraries can promote pathfinders accessible both in online and print formats in a variety of ways.

Traditional Promotion

Pathfinders can be accessed in the traditional way—at the library. Printed versions can be displayed at the reference desk or within the collection in the appropriate subject area or even mailed to potential customers, such as groups or agencies with an information need. Printed pathfinders should always include a note indicating the Web address of where the most current version is available.

Online Promotion

There are several methods for improving the awareness of online pathfinders, including linking from one related pathfinder to another. This category may be best displayed at the end of the pathfinder as "related pathfinders." For example, the pathfinder on how to start a business can link to pathfinders that cover trademarks or company information.

Links to pathfinders can also be e-mailed to faculty, staff, students, researchers, and interested organizations and groups. You must determine a pathfinder's intended audience, in most cases, by individual topic. For example, if the pathfinder topic is genealogy, then the most likely audience is the local genealogy libraries and senior organizations. Normally an e-mail to that organization with the appropriate link, description, and perhaps even a graphic makes it easier for other Web sites to add the link to their site. This takes time, but can really pay off. Web statistics software may even be able to generate a list of the top referral Web sites. This offers a view of how Web site users are reaching your pathfinders.

Cooperative Efforts and Commercial Ventures

Due to the amount of time and the repetitiveness of the inquiry some library organizations have published pathfinders to share among member libraries. Individual libraries have traditionally authored pathfinders, which assumes that the library owns or provides access to all of the material listed in its pathfinder. Cooperative and commercially-written pathfinders may include some materials not owned by the library; however, these pathfinders may be just as valuable for researchers (Miller, 2000). For example, a researcher may find material that he or she can borrow through interlibrary loan or purchase commercially. Although libraries do not own all the materials contained in databases such as WorldCat (OCLC), Novelist (Ebsco Publishing), or Books in Print (R. R. Bowker), researchers still find these resources useful in identifying resources about which they would have otherwise been unaware.

Model Library Project

The Model Library Project, an initiative at Massachusetts Institute of Technology's Baker Engineering Library during the early 1970s, sought to satisfy the user's informational questions at their point-of-need (Dunsmore, 2002). After a failed attempt to produce the guides cooperatively with other libraries, the Model Library Project negotiated with Addison-Wesley Publishing to market and distribute the guides from 1972 to 1975. Both efforts revealed that libraries did not have the time to develop guides reciprocally, nor were they willing to purchase commercial pathfinders that did not match local collections (Dunsmore, 2002). In an electronic environment where files can be easily updated and posted to the Web and free Web-based materials are accessible to everyone with access to the Internet, the concept of collaboratively produced pathfinders becomes more appealing than print pathfinders shared among libraries in the Model Library Project. Interlibrary loan is much more streamlined, as well. Libraries with small or no print collections (like online colleges), who choose to include materials not owned in their catalogs, should include links to services such as interlibrary loan and document delivery services.

SIRSI Rooms (Sirsi Corporation)

www.sirsi.com/Sirsiproducts/rooms.html

Sirsi Corporation created software that allows libraries to build subject-specific content on Web pages, which incorporate various formatted materials including licensed databases. Libraries can also use templates or customize off-the-shelf Rooms Blueprints, or by creating their own unique pages through the Rooms interface.

OCLC Pathfinders

www.oclc.org/connexion/about/features/pathfinders/

OCLC's cataloging service, Connexion, offers the ability to create online pathfinders or use another libraries saved pathfinders. You may also export and save them offline for use on the library Web site.

Research Guide

http://researchguide.sourceforge.net

Developed by a librarian and student at the University of Michigan, ResearchGuide is an open-source free tool that allows staff to automate the creation and maintenance of subject guides for libraries. The software also generates a list of subject specialist that coordinate each page.

Cooperatively or commercially produced pathfinders can be useful if you can easily adapt them to fit local collections and services. Librarians can localize pathfinders by adding hypertext links from book titles into the library catalog, licensed databases, and/or to local agencies and organizations.

2
Writing Pathfinders

Preparing pathfinders includes time for identifying, evaluating, and organizing resources, and writing annotations. Additional time is necessary for editing, review, and link checking. Finding aids used to locate quality resources are available in print and online. During the selection process librarians must locate, analyze, and synthesize resources. The Internet serves as a resource for many online materials, and, in fact, due to the plethora of information available, may even increase the amount of time spent sorting through resources. Although used as a vehicle to disseminate information quickly, efficiently, and cost-effectively, the Internet does not necessarily decrease the time it takes to prepare the guides.

Libraries that have limited print collections may want to rely more heavily on free Web sites and nearby print collections depending on reciprocal borrowing arrangements these libraries share. Each library will approach these issues differently; however, here are some questions that may arise:

1. *What is the length of the pathfinder? Should a minimum or maximum length be set?*

   Because pathfinders are not meant to be exhaustive and merely serve as an introduction to basic sources, they should be kept to approximately four screens or two printed pages in length.

2. *What is the maximum number of items listed in each bibliographic unit and/or category?*

   Obtaining a balance of resources in each category is preferable; however, including the most relevant materials based on authority, accuracy, currency, and maintaining a balance of viewpoints is more important than adhering to quantitative methods of inclusion.

3. *Web sites may include an online directory of organizations or an online dictionary. Should these items be listed under Web sites or in other categories that list print materials?*

   Base the decision on content. If a Web site identifies with an existing category, then place it there. This issue is covered in further detail in Chapter 3.

When questions arise in creating the pathfinders, reflect on the customers needs and expectations. Think about how students and researchers will use the pathfinder and what best suits their needs. In preparing pathfinders, each library ultimately determines what is most suitable for the topic covered and its audience and collection.

Selecting Topics

Librarians can begin to develop a list of the most frequently asked questions and requested topics to determine what pathfinders to create. In a multibranch system elicit the input from frontline staff who work directly with the library's patrons. Librarians should collaborate with schools and universities to ensure that the topics are relevant to class assignments. In most cases it will be librarians that create and maintain these pages so it is important to involve them early on.

Sometimes an event occurs that prompts the creation of a pathfinder to satisfy an informational need (see Figure 2-1). For instance, the terrorist attacks on September 11, 2001 prompted the creation of many pathfinders. Librarians rushed to create pathfinders in the wake of this tragedy. An archive of selected sites created for this purpose is maintained at a special page at the Internet Archive (*http://web.archive.org/collections/sep11.html*), including library Web pages.

Libraries may also create two pathfinders that serve the differing informational needs of adults and children. Although the topic is the same, the resources, language, and format of the pathfinder are presented entirely differently. For example, the Indianapolis Marion County Public Library maintains a pathfinder for children titled, *Elections and Voting* (*http://infozone.imcpl.org/kids_path_elections.htm*), in addition to a subject guide for adults titled, *Elections* (*www.imcpl.lib.in.us/ss_elections.htm*) (see Figure 2-2).

Gathering Content

When creating pathfinders tailored to the library's collection, the first place to start, of course, is the library catalog. However, since pathfinders are meant to be selective, how do you determine what are the most appropriate titles to include? Researchers expect that librarians have filtered out the mediocre titles and what are listed in a pathfinder are the most relevant resources. The tools librarians use for collection development can also be used to identify worthwhile resources for inclusion into the pathfinder. The reference guides below provide descriptions and evaluations of many print and titles.

American Reference Books Annual (ARBA), Volume 34. Westport, CT: Libraries Unlimited, 2003.
www.arbaonline.com.
Includes over 1,800 descriptive and evaluative entries of recent reference publications. Reviews are written by subject experts on publications from more than 300 publishers in 500 subject areas. Look for a new series titled *ARBA In-Depth*, which covers children's and young adult titles, health and medicine, economics and business, and philosophy and religion.

Best Free Reference Web Sites
www.ala.org/rusa/mars/
Annual list of useful Web sites selected by Machine-Assisted Reference Section (MARS) of the Reference and User Services Association (RUSA) of ALA.

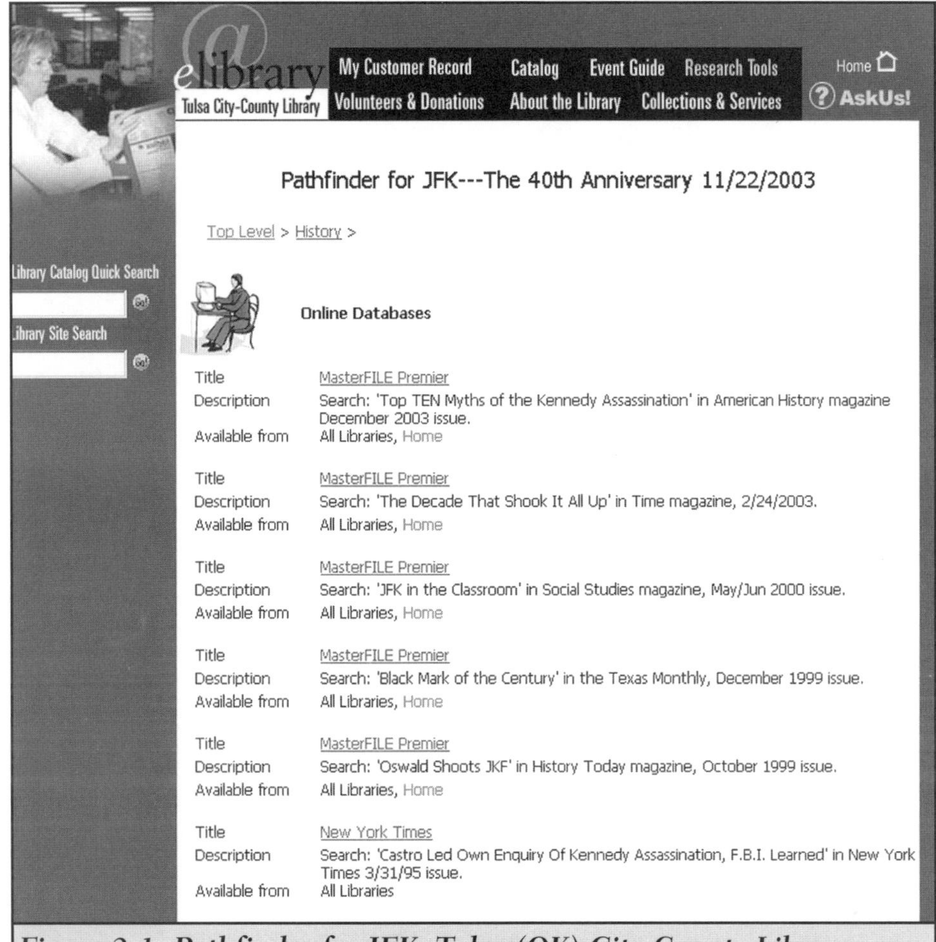

Figure 2-1. Pathfinder for JFK, Tulsa (OK) City-County Library
Pathfinder for JFK, Tulsa City-County Library, www.tulsalibrary.org/research/ORC_Topic.asp?ID=192.

Best Information on the Net
St. Ambrose University, O'Keefe Library
http://library.sau.edu/bestinfo/
Offers annotated links to great Web sites, organized by major, paper topics, and alphabetical indexes.

Booklist's Editor's Choice
www.ala.org/ala/booklist/editorschoice/booklisteditors.htm
Lists top titles in adult books, books for youth, media, and reference. Includes publication information and annotations.

Guide to Reference Sources, 12th Edition. Robert Kieft, General Editor. Chicago: ALA Editions. Forthcoming.
This standard reference tool began in 1902 as Alice Bertha Kroeger's work, *Guide to the Study and Use of Reference Books: A Manual for Librarians, Teachers and Students,* transformed to the *Guide to Reference Books,* by editors Mudge, Winchell, Sheehy, and Balay, and will take a new title and edition under the editorial direction of Robert Kieft. Provides a bibliography to the most standard reference books and online sources. For publication updates see *www.haverford.edu/library/grb.*

Elections

Home > eReference > Subject Pathfinders >

New on our Site | Locations & Hours | Online Databases | Email the Webmaster | Search our Site

Printable Version

Description

The Indianapolis-Marion County Public Library has resources that inform people about the federal, state, and local elections. For more information about government, politics, and elections, visit the Library's Local Government and Government and Politics subject pages.

FAQ | Election Offices | Books & Other Library Materials | Online Databases | Web Sites

Frequently Asked Questions About Elections

When will the next election be held?

The next election, a general election, will be held **Tuesday, November 2, 2004**.

The Indiana Election Division issues the *2004 Indiana Election Calendar*, which includes deadlines for candidate filings, voter registration, and absentee ballots. The Marion County Election Board maintains a calendar that includes meetings and deadlines.

What offices are up for election in 2004?

In 2004, Marion County voters will elect the following:

Federal Offices

President and ***Vice-President*** of the U.S.

U.S. Senate: one seat (Evan Bayh's)

U.S. House of Representatives: Districts **4**, **5**, and **7**

Elections & Voting

infoZone > Pathfinders >

Kids' Catalog | FindIt | Pathfinders | Inspire Kids | Activities | E-mail Ozo | About

- This pathfinder is a guide to factual books, stories, web sites and more about voting and elections. You can use it to learn about the rules that elections follow, about famous elections from the past and also keep up with news about current elections.

Books | Audio Visual | Web Sites
Related Catalog Searches | Related Subject Pathfinders for Kids

---

jP C9475du
Cronin, Doreen
Duck for President
Summary: When Duck gets tired of working for Farmer Brown, his political ambition eventually leads to his being elected President.

jP K9387m
Krosoczka, Jarrett
Max for President
Summary: Max and Kelly both want to win the election for class president, but when one of them loses, the winner finds a way to make the loser feel better.

Figure 2-2. Elections for Children and Adults, Indianapolis-Marion (IN) County Library

The Internet Scout Project
http://scout.wisc.edu
Continuously published since 1994. Weekly publication alerting researchers to valuable resources available on the Web.

Internet Public Library
www.ipl.org
Thousands of links to authoritative Web sites. Also includes special collections, such as Associations on the Net, Presidents of the United States, and Literary Criticism.

KidsClick!
http://sunsite.berkeley.edu/KidsClick!/
Created by a group of librarians at the Ramapo Catskill (NY) Library System to identify and make available to young people valuable and age appropriate Web sites.

Librarian's Index to the Internet
www.lii.org
Subject directory of Internet sites that assists librarians in identifying quality information. Provides descriptions of each Web site.

Neat New Stuff on the Net Marylaine Block
http://marylaine.com/neatnew.html
Provides listings of useful new Web sites each week.

Reader's Adviser. 14th edition. Ed. Marion Sader. Westport, CT: Libraries Unlimited. 1994. ISBN: 0835233200
Features hundreds of authors and thousands of works including critical reviews. Includes title, name, and subject indexes in every volume. An annotated guide to the best in print, literature, biographies, dictionaries, encyclopedias, Bibles, classics, drama, poetry, fiction, science, philosophy, travel, history. Includes the following volumes:

- Vol. 1 *The Best Reference Works, British Literature* (ISBN: 0835233219)
- Vol. 2 *The Best in World Literature* (ISBN: 0835233227)
- Vol. 3 *The Best in Social Sciences, History and the Arts* (ISBN: 0835233235)
- Vol. 4 *The Best in Philosophy and Religion* (ISBN: 0835233243)
- Vol. 5 *The Best in Science, Technology* (ISBN: 0835233251)
- Vol. 6 *Indexes* (ISBN: 083523326X)

Recommended Reference Books for Small and Medium-Sized Libraries and Media Centers. Westport, CT: Libraries Unlimited, 2003. ISBN: 1591580552
This annual volume assists libraries in the selection of materials and reference works. Includes approximately 500 brief reviews from the current edition of *American Reference Books Annual* and are selected as they are appropriate for inclusion in small or midsize library collections. Reviews are organized by discipline.

Subject Bibliographies, United States Printing Office (USGPO)
http://bookstore.gpo.gov/sb/sale180.html
Offers approximately 150 subject bibliographies that are used to categorize the publications, subscriptions, and electronic products for sale by the Superintendent of Documents, United States The lists are not selective but offer listings of materials currently available through the USGPO.

Walford's Guide to Reference Materials. Volume 1: Science and Technology. 8th Edition. Ed. Marilyn Mullay and Priscilla Schlicke. London: The Library Association, 1999. ISBN: 185604341X
Contains evaluative annotations to reference sources in each subject area: mathematics; astronomy and surveying; physics; chemistry; earth sciences; paleontology; anthropology; biology; natural history; botany; zoology; patents and interventions; medicine; engineering; transport vehicles; agriculture and livestock; household management; communication; chemical industry; manufactures; industries, trades and crafts; and the building industry.

Walford's Guide to Reference Material. Volume 2: Social and Historical Sciences, Philosophy and Religion. 8th Edition. Ed. Alan Day and Michael J. Walsh. London: The Library Association, 2000. ISBN: 185604369X
Contents include philosophy, psychology, religion, social sciences, sociology, statistics, politics, economics, labor and employment, land and property, business organizations, finance and banking, economic surveys, economic policies and controls, trade and commerce, business and management, law, public administration, social services and welfare, education, customs and traditions, geography, biography, history. Walford's series title that contains other subject areas (forthcoming Vol. 1, Science, medicine, technology, and industry).

Additional Elements

In addition to bibliographic units and categories such as books, Web sites, and subject headings (see pages xvii–xxi for a complete list), libraries may want to consider other categories not traditionally included in pathfinders. Other categories outside the scope of formats and topics include frequently asked questions, how to evaluate information, and links to related academic departments and course pages. Even so, some categories may work well in some pathfinders but may not be appropriate for others. For example, an academic library would probably want to include citation styles in each pathfinder and a public library may want to include a link to the event calendar if they offer programs about specific topics, such as job hunting or patent searching. Allow for some flexibility within each pathfinder. Focus not on standardization, but meeting the researcher's informational needs. RUSA's *Guidelines for the Preparation of a Bibliography* states that "the compiler should include all available bibliographic units within the subject. A bibliographic unit is an entity in a bibliography: book, journal articles, reports, manuscripts, sound and video recordings, individual Web pages and/or entire Web sites, computer programs or printouts, films, charts, etc."

Linking to Library Services

Libraries may want to include a link or icon, which highlights some of the services it provides such as interlibrary loan, online or telephone reference services, or other services. For example, in a pathfinder for senior services the library may wish to include a link to a page describing a books-by-mail service.

Writing Annotations

An annotation is a summary and/or evaluation of the sources listed in a bibliography. For purposes of a pathfinder researchers would normally conclude that the source was included because it is reliable, unbiased, objective, accurate, and authoritative, and therefore, an evaluative statement may not be necessary. After reading the annotation students should be able to determine whether or not the resource would be helpful to their research. The summary should clearly distinguish one resource from another. RUSA's suggested guidelines for annotations and notes state that they should be succinct, informative, and on a level appropriate for its audience.

Editorial Guidelines

For consistency in presentation of material, pathfinders should follow a defined set of guidelines including grammar, design, and format. Since pathfinders may be authored by many contributors it would be helpful to determine editorial style guides that should be followed, such as the *Chicago Manual of Style* or the *Publication Manual of the American Psychological Association*. In addition to a style manual, some libraries have designated a specific dictionary for general use and may also create a list of frequently used Web-related terms, generated so that staff follows certain standards (e.g., on-line or online, disk or disc, web page or Web page). Additionally, guidelines on the placement of each resource within a category may be helpful. For example, a directory of businesses available on a Web site can be categorized as either unit, Web site, or directory. Guidelines can ensure more consistency among pathfinders that are produced by many authors.

Maintenance

Pathfinders published to the Web can be readily updated and continuously revised. The challenge is that the pathfinder is never fixed in time like its print counterpart, and thus, the pathfinder is virtually never finished. It must be maintained on an ongoing basis so that it remains relevant, old sources are weeded, new ones added, and dead links are either remedied or deleted.

Although link checkers provide an automated way to check for inactive links they must be checked manually. If a domain name license expires a new owner may radically change the content of the site. Librarians should also determine if any site changes make the site ineligible for inclusion based on established selection criteria. For example, if a site that was once free becomes subscription-based or plug-ins make accessing content too cumbersome, it may no longer meet your criteria. Additionally, if the pathfinder links to a specific electronic journal in an online database that is subsequently dropped by the vendor, the librarian may want to pull the listing entirely or find a comparable journal. Many times aggregators drop and add journal titles without notification to subscribers. Librarians should also check links to databases that require authentication for remote users. The following Web sites offer free link checking:

Doctor HTML
www.doctor-html.com/RxHTML/

W3C's HTML Validator
http://validator.w3.org

Web Design Group's HTML Validator
www.htmlhelp.com/tools/validator/

The easiest pathfinders to maintain are the ones that display dynamic content using information that originates from a database. A library can create a database-backed pathfinder collection by establishing standardized fields, such as title, introduction, books, Web sites, and subject headings; however, you must incorporate a way to add new subject headings for specific pathfinders.

Additionally, content from a pathfinder can be generated externally from another internal database source. Take, for example, a pathfinder about computers that includes all of the library's computer classes in an area of the pathfinder. As a staff member enters computer classes into an events database, a link on the computer pathfinder is sending a command to the database to display all of the upcoming computer classes. Regardless of whether or not the pathfinder is designed to generate dynamic content it will, just like any other Web page, need to be revised at some point. Academic and school libraries may want to update their pathfinders just prior to the new school year.

In order to continue to maintain the collection of pathfinders, libraries use the following guidelines:

1. An assigned staff tends to the pathfinder on a regular basis.
2. Staff will determine how frequently they will update the pathfinder.
3. The pathfinder is made available on a staging or development server so authoring staff can review it prior to posting it to the public.

3
Presenting Pathfinders on the Web

Publishing pathfinders on the Web provides librarians with as many opportunities as it does challenges. The most prevalent feature of Web-based pathfinders is hypertext links, because they provide researchers with access to resources, something that cannot be accomplished through print pathfinders. In addition, advantages include cost efficiency, 24-hour accessibility for remote users, and ability to easily update them.

Organization of Resources

Pathfinders can be organized by format, subject, or a combination of both. There are no rules on how to author pathfinders; however suggested authoring guidelines are available in a variety of the literature. A quick review of several library Web sites demonstrates the variety of approaches that librarians have taken. Guidelines for authoring are difficult to standardize and will continue to vary as they do now. The following RUSA guidelines (2001) address the organization of bibliographies:

1. The organization of material should be suitable for the subject and the targeted users.
2. The main arrangement should make it possible to use the bibliography from at least one approach without consulting the index.
3. Multiple means of access should be provided if appropriate. Means of access include both the meaningful arrangement of materials and the available indexes to those materials.
4. The scheme for a classified bibliography should be logical and easy for users to understand.
5. Bibliographies published on the World Wide Web should make use of recognized navigation features and other sound principles relating to layout and file size.

Many pathfinders include bibliographic units by format. Perhaps this was done to satisfy teachers' requirements that students find a certain amount of materials in specific formats (e.g., three magazine articles, two books, one encyclopedia). Dunsmore explains that the traditional format organization of pathfinders

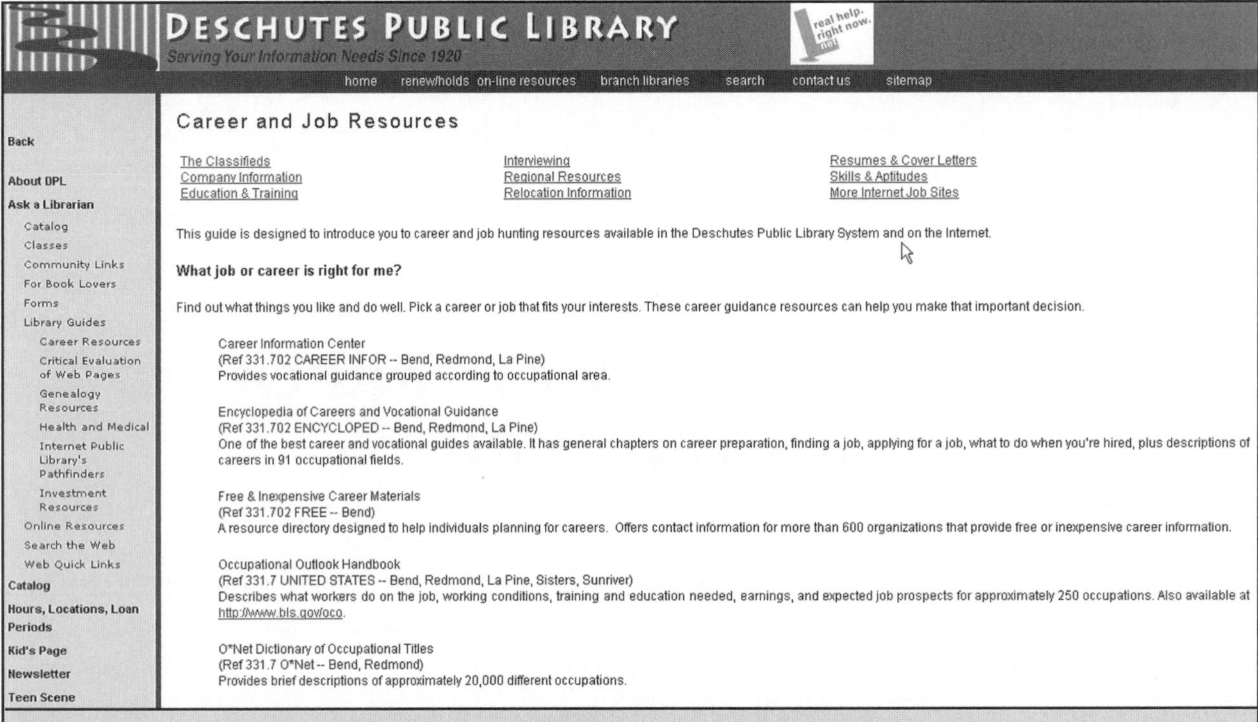

Figure 3-1. Inquiry-Based Pathfinders, Deschutes (OR) Public Library

An inquiry-based approached to guided searching, Deschutes Public Library, www.dpls.lib.or.us/Page.asp ?NavID=138

was superseded by organizing modern, electronic guides according to the topical subsets of information. Some guides, such as the Career and Job Resources pathfinder produced by Deschutes (OR) Public Library in Figure 3-1 are organized by frequently asked questions. This inquiry-based approach is consistent with the questions librarians may receive when working directly with the patron.

When applying format-oriented categories it is clear that the category of "Web sites" offers some challenges. For example, the U.S. Chamber of Commerce Web site provides a directory of members. This resource can be categorized under directories, organizations, and Web sites. "Websites that were listed within the various guides were not consistently grouped into this 'Websites' theme, as many pathfinders grouped them under other categories, such as, associations, governmental statistics and non-governmental organization" (Dunsmore, 2002: 146).

Usability

Usability refers to the ease with which a person is able to navigate and utilize a Web page. According to Jakob Nielsen (2003), the usability expert, usability has five quality components:

1. *Learnability:* How easy is it for users to accomplish basic tasks the first time they encounter the design of a Web page?
2. *Efficiency:* Once users have learned the design, how quickly can they perform tasks?

3. *Memorability:* When users return to the design after a period of not using it, how easily can they reestablish proficiency?
4. *Errors:* How many errors do users make, how severe are these errors, and how easily can they recover from the errors?
5. *Satisfaction:* How pleasant is it to use the design?

In order to achieve their maximum usability, authoring librarians must consider a pathfinder's readability, navigation, hypertext links, and style guides.

Readability

Researchers must be able to read easily and scan through the pathfinder in search of relevant sources of information. Online users prefer to skim and read small amounts of text online, unless they have reached their informational destination. Librarians should ask the following questions: Is the text clearly written for its intended audience? Is the text kept to a minimum and easier to read by chunking information or using bulleted lists? Text on a resource list like a pathfinder is best kept to a minimum.

Navigation

Navigation refers to the mechanisms that users must utilize to move from an index page to a pathfinder and throughout the pathfinder. Such movements include linking to other resources, returning back to the pathfinder, and moving throughout the main library Web site. Questions librarians should ask include: Is the user able to navigate easily through the document from the table of contents to specific sections within the document and back to the top of the page and the Web site's main navigation bar? Can the researcher easily return back to the main index of all pathfinders or to the library's Web page? If the library offers so many pathfinders and browsing through topics becomes unwieldy, do they provide a search engine?

Images

Images can greatly improve the design of Web pages; however, since most visitors to library Web sites are information seekers, images do not necessarily enhance their experience. Users will not tolerate lengthy waits for pages to load. Figure 3-2 illustrates the practical use of images.

Printer-Friendly Pages

A printer-friendly version of a Web page excludes images and frames, and provides all resources in one document for easy printing. It may be used by the researcher who wants to browse the stacks with the document in hand or the librarian who wants to use it as a handout for a visiting class. Although certain scripts and cascading style sheets can automate the process of creating such pages, librarians should determine whether or not they want to make the URL visible to the user. Authors would have to make the Web site addresses visible so that when the page is printed the URLs can be seen. For some of the more lengthy URLs this can become quite cumbersome to present. If librarians believe that the online pathfinders will be used offline then they may want to consider creating printer-friendly pages, either manually or through an automated script.

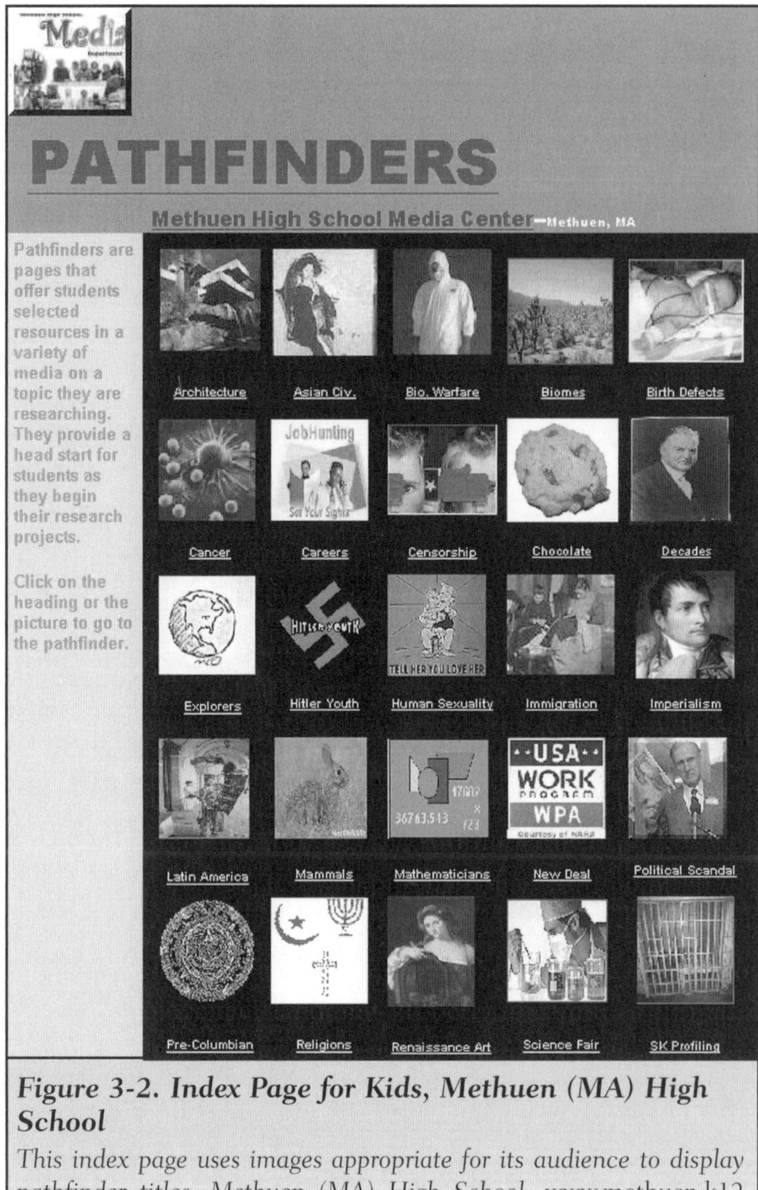

Figure 3-2. Index Page for Kids, Methuen (MA) High School

This index page uses images appropriate for its audience to display pathfinder titles. Methuen (MA) High School, www.methuen.k12.ma.us/pathfinders.

Templates

Templates can be created to ensure that page elements appear consistently throughout the collection of pathfinders. In addition to bibliographic units (e.g., books, magazines, newspapers), a pathfinder may also include links to online help, the library logo, contact information, URL, date created, date revised, citation guides, Web site main navigation bar, and copyright notice. A wide range of possible page elements appear. Templates can also be produced that serve a specific audience, for example, children. Children may prefer larger text and a lot of images with splashy colors. Adults, on the other hand, may prefer a design unencumbered by graphics.

Cascading Style Sheets (CSS)

Cascading style sheets allows Web designers to apply certain design elements such as type (font) faces, colors, sizes, and background colors to a structural Web document. Not only does the use of style sheets offer a consistent design to the pathfinders, but it also provides the ability to change it quickly by altering elements in the style sheet. This change is done once without having to touch each Web page. Style sheets provide a wonderful maintenance benefit. Furthermore, visitors can apply their own style sheets for those with limited or low vision. For a current review of cascading style sheets and to learn about the newest standards, visit the World Wide Web Consortium Web site (Available: *www.w3.org/Style/CSS*).

HTML/XHTML

HTML (Hypertext Markup Language) has traditionally been the scripting language used to produce Web pages; however, a newer specification for Web page authoring, XHTML (Extensible Hypertext Markup Language), has recently been adopted by the World Wide Web Consortium, an organization whose mission it is to develop common Web protocol and guarantee the Internet's interoperability. Pathfinders should follow current Web authoring guidelines to ensure that they are made available to the widest possible audience. For more information on the current versions and latest development on Web publication see the World Wide Web Consortium's page on hypertext markup language (Available: *www.w3.org/MarkUp*).

Hypertext Links

The use of hypertext enables visitors to get more information about a particular resource by clicking on its title. Hypertext links can be created from the pathfinder into the catalog by title, author, subject heading, ISBN, or other identifiers. Additionally, links to full-text magazine articles allow greater usability for Web users. As more print publications are transferred to the Web and libraries purchase an increasing amount of electronic resources, the ability of the pathfinder to meet information needs with a click of the mouse increases the effectiveness of the pathfinder.

The ability to link to resources has transformed the pathfinder, but it also has some challenges inherent to its technology, which impact its structure, presentation, and usability. Libraries may determine that, although cumbersome, URLs should be listed alongside the title of each target. As discussed above, including the full URL ensures that printed pathfinders remain useful when they are printed. However, to make this determination, libraries should ask themselves if the majority of customers will use the pathfinder online or offline.

Structural Links

Structural links meet the navigation user's need to move throughout the pathfinder. This can be accomplished by placing named anchors within the Web page using the following HTML tag:

```
<a name="circulating"></a>
```

Authors place these tags throughout the document like placing bookmarks. Each anchor should be named appropriately. When linking to that anchor the tag should look like this:

```
<a href="#circulating">Circulating Materials</a>
```

A table of contents that represent each bibliographic unit should be available at the top or the left side of the document. At the end of each bibliographic unit it is important to create a link that brings users back to the table of contents. This allows them to navigate throughout the pathfinder easily without having to scroll continually to the top of the document.

Embedded Links

Embedded links refers to the traditional links that Web users see as underlined and in blue text. They intuitively know that if they click on the text they will be forwarded to more information about whatever the text implies.

These links will bring visitors to a specific page of a Web site beyond the home page. Even more valuable is the ability to link to a database—either a dynamic search or a specific entry. For example, creating a link that executes a subject search of the catalog will deliver dynamic content to the user. As the catalog is updated with new materials in that subject, it retrieves a new set of results. Each time a visitor selects that link the search is executed.

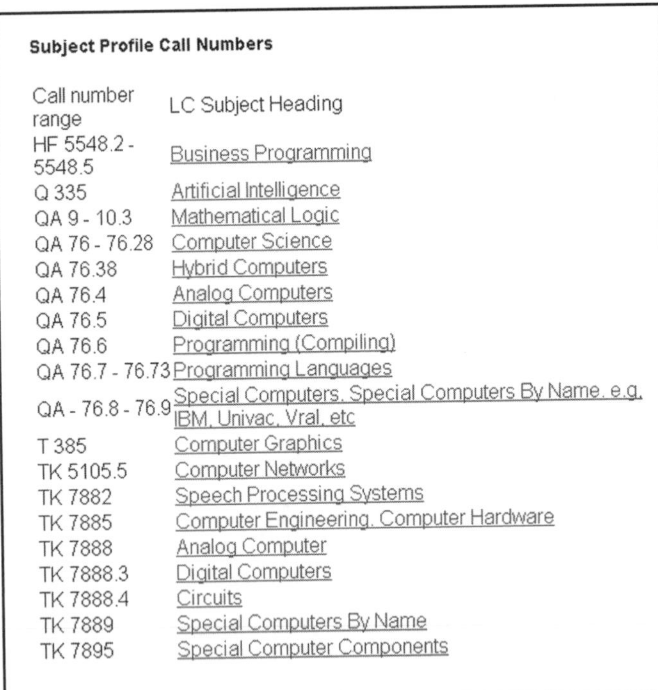

Subject Profile Call Numbers

Call number range	LC Subject Heading
HF 5548.2 - 5548.5	Business Programming
Q 335	Artificial Intelligence
QA 9 - 10.3	Mathematical Logic
QA 76 - 76.28	Computer Science
QA 76.38	Hybrid Computers
QA 76.4	Analog Computers
QA 76.5	Digital Computers
QA 76.6	Programming (Compiling)
QA 76.7 - 76.73	Programming Languages
QA - 76.8 - 76.9	Special Computers, Special Computers By Name, e.g. IBM, Univac, Vral, etc
T 385	Computer Graphics
TK 5105.5	Computer Networks
TK 7882	Speech Processing Systems
TK 7885	Computer Engineering, Computer Hardware
TK 7888	Analog Computer
TK 7888.3	Digital Computers
TK 7888.4	Circuits
TK 7889	Special Computers By Name
TK 7895	Special Computer Components

Figure 3-3. Dynamic Link to Catalog, Auburn University (AL)

Embedded links allow researchers to move from subject headings to a search in the library catalog. Auburn University Libraries, www.lib.auburn.edu/scitech/resguide/computer/compusci.html.

Although Web publishing standards do not recommend hypertext links opening up new browsers it is important that the pathfinder remain on the desktop so that users can refer back to it. If links do not open up into a new browser, then patrons may lose their point of reference. Authoring librarians should also realize that, although Internet technology allows us to create very complex hypertext documents, the researcher's main task is to gather materials for their research project by using the pathfinder to assist them in locating those materials. Again, keeping the intent of the document and needs of the customer in mind when designing each pathfinder helps ensure it remains a valuable research tool.

Linking to Licensed Databases

An image or notation should be placed nearby links to licensed databases so that researchers know that they may need their library card to access the listed resource. If the resource is accessible only from within the library, then a notation or icon should alert the researcher to that, as well. Additionally, once a researcher has typed in their library barcode and is authenticated to use licensed databases, he or she should not be required to input a library card number again for subsequent searches into other fee-based database.

Deep Linking into a Database

When a Web site links to another site's page, but not the home page, it is considered deep linking. A very powerful method of bringing researchers relevant information is the ability to link into a database, either one the library subscribes to or a public domain databases. These types of database are often referred to as the invisible Web, because the information contained in them is quite often not retrieved by search engines. The three types of links that one may encounter are temporary links, links that require authentication (also those that are proprietary may expire after a stated time period), and those that work indefinitely.

Some links may expire after a certain length of time so it may need to be tested for continued access. Some vendors of licensed subscription databases, such as the Gale Group, offer the use of a persistent URL. The Gale Group calls this functionality an InfoMark, which can be incorporated nicely into a pathfinder. It allows researchers to copy and paste the URL of an article into their documents so that others can access the same article or repeat a search. Only visitors with a library card can access this information.

Deep linking into a public domain database can bring very successful results; however, links must be tested to make sure they do not time out. For example this URL will retrieve Congressional Record H12928 dated December 15, 2003: *http://frWebgate.access.gpo.gov/cgi-bin/getpage.cgi?dbname=2003_record&page=H12928&position=all.* Linking to temporary links should be avoided. For example, the following URL will not work: http*://thomas.loc.gov/cgi-bin/query/D?c108:11:./temp/~c1081G58Bn::*

Exemplary Pathfinders

The following screen shots of library pathfinders provide insight into the design and structure of exemplary subject guides. Each page highlights exemplary design elements and also challenges that users face in using the guide.

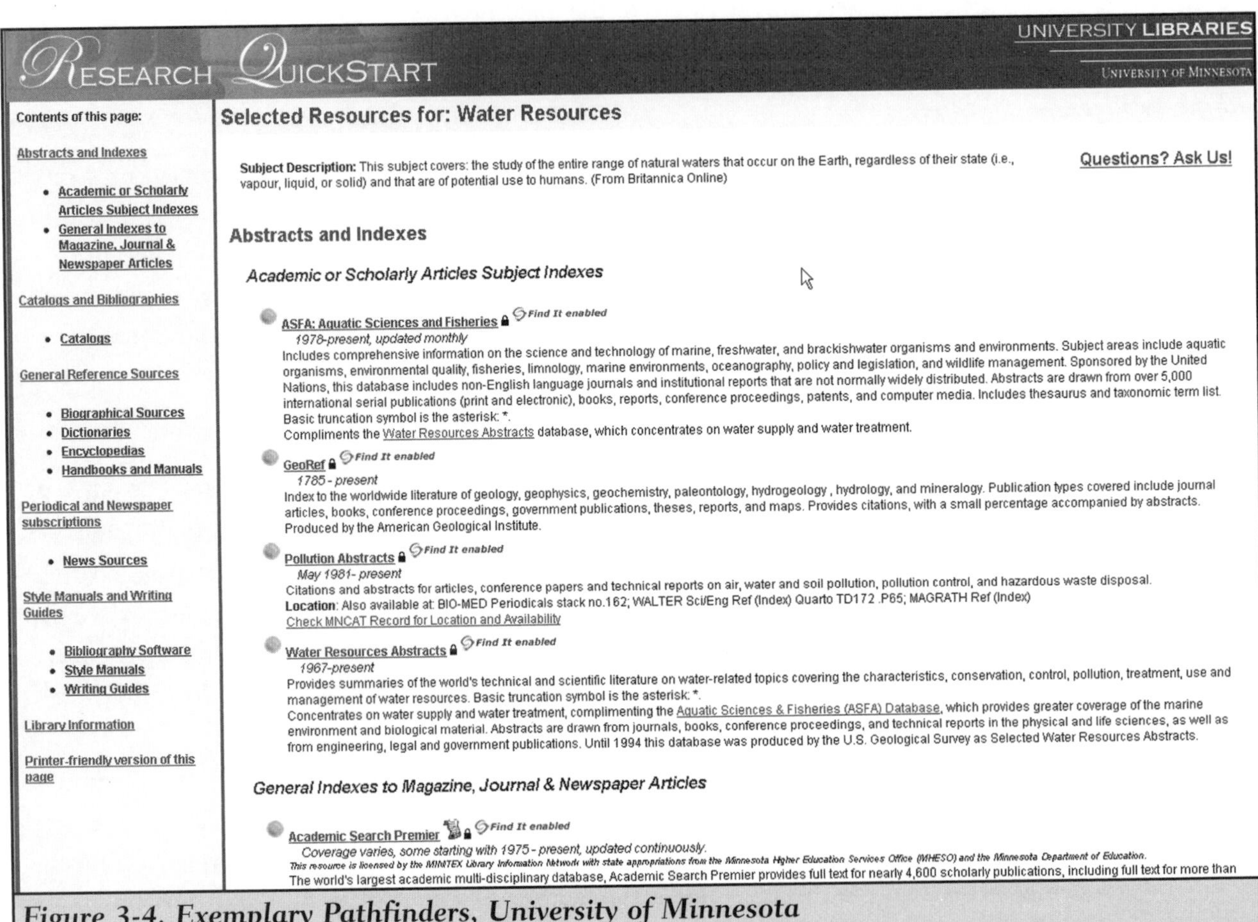

Research QuickStart

UNIVERSITY LIBRARIES

University of Minnesota

Contents of this page:

Abstracts and Indexes

- Academic or Scholarly Articles Subject Indexes
- General Indexes to Magazine, Journal & Newspaper Articles

Catalogs and Bibliographies

- Catalogs

General Reference Sources

- Biographical Sources
- Dictionaries
- Encyclopedias
- Handbooks and Manuals

Periodical and Newspaper subscriptions

- News Sources

Style Manuals and Writing Guides

- Bibliography Software
- Style Manuals
- Writing Guides

Library Information

Printer-friendly version of this page

Selected Resources for: Water Resources

Subject Description: This subject covers: the study of the entire range of natural waters that occur on the Earth, regardless of their state (i.e., vapour, liquid, or solid) and that are of potential use to humans. (From Britannica Online)

Questions? Ask Us!

Abstracts and Indexes

Academic or Scholarly Articles Subject Indexes

ASFA: Aquatic Sciences and Fisheries Find It enabled
1978-present, updated monthly
Includes comprehensive information on the science and technology of marine, freshwater, and brackishwater organisms and environments. Subject areas include aquatic organisms, environmental quality, fisheries, limnology, marine environments, oceanography, policy and legislation, and wildlife management. Sponsored by the United Nations, this database includes non-English language journals and institutional reports that are not normally widely distributed. Abstracts are drawn from over 5,000 international serial publications (print and electronic), books, reports, conference proceedings, patents, and computer media. Includes thesaurus and taxonomic term list. Basic truncation symbol is the asterisk: *.
Compliments the Water Resources Abstracts database, which concentrates on water supply and water treatment.

GeoRef Find It enabled
1785 - present
Index to the worldwide literature of geology, geophysics, geochemistry, paleontology, hydrogeology, hydrology, and mineralogy. Publication types covered include journal articles, books, conference proceedings, government publications, theses, reports, and maps. Provides citations, with a small percentage accompanied by abstracts. Produced by the American Geological Institute.

Pollution Abstracts Find It enabled
May 1981- present
Citations and abstracts for articles, conference papers and technical reports on air, water and soil pollution, pollution control, and hazardous waste disposal.
Location: Also available at: BIO-MED Periodicals stack no.162; WALTER Sci/Eng Ref (Index) Quarto TD172 .P65; MAGRATH Ref (Index)
Check MNCAT Record for Location and Availability

Water Resources Abstracts Find It enabled
1967-present
Provides summaries of the world's technical and scientific literature on water-related topics covering the characteristics, conservation, control, pollution, treatment, use and management of water resources. Basic truncation symbol is the asterisk: *.
Concentrates on water supply and water treatment, complimenting the Aquatic Sciences & Fisheries (ASFA) Database, which provides greater coverage of the marine environment and biological material. Abstracts are drawn from journals, books, conference proceedings, and technical reports in the physical and life sciences, as well as from engineering, legal and government publications. Until 1994 this database was produced by the U.S. Geological Survey as Selected Water Resources Abstracts.

General Indexes to Magazine, Journal & Newspaper Articles

Academic Search Premier Find It enabled
Coverage varies, some starting with 1975 - present, updated continuously.
This resource is licensed by the MINITEX Library Information Network with state appropriations from the Minnesota Higher Education Services Office (MHESO) and the Minnesota Department of Education.
The world's largest academic multi-disciplinary database, Academic Search Premier provides full text for nearly 4,600 scholarly publications, including full text for more than

Figure 3-4. Exemplary Pathfinders, University of Minnesota

University of Minnesota, University Libraries
Research QuickStart
http://research.lib.umn.edu

1. All pathfinders at this Web site are titled Research QuickStart, which most students will understand.
2. This logo is a link that brings you to the main university homepage and the library's homepage.
3. The table of contents on the left side of the page corresponds to the bibliographic units within the pathfinder.
4. A brief description indicates its source.
5. The site invites students, who would like more information about a database, to inquire further.

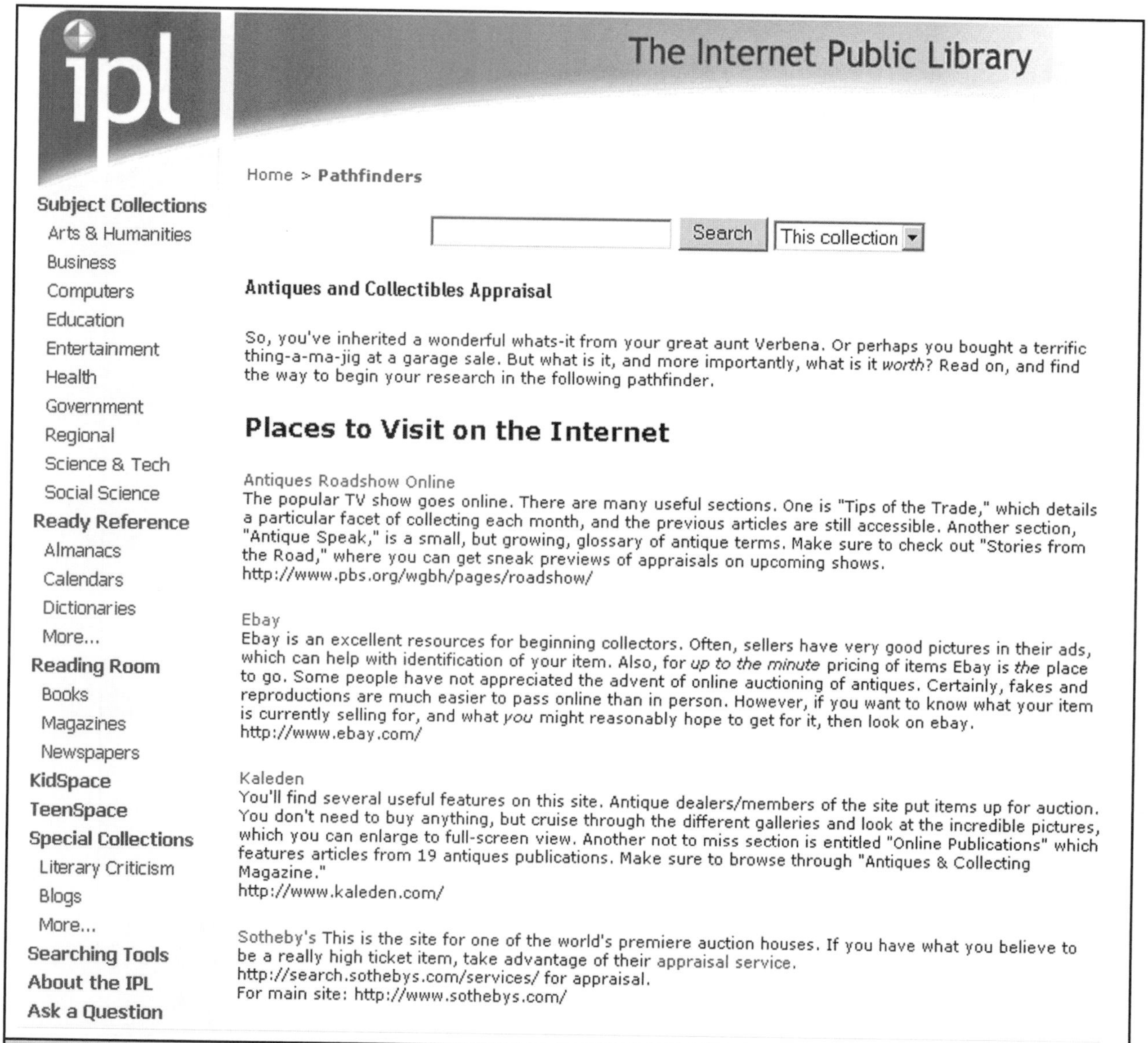

ipl

The Internet Public Library

Home > **Pathfinders**

Search | This collection

Subject Collections
- Arts & Humanities
- Business
- Computers
- Education
- Entertainment
- Health
- Government
- Regional
- Science & Tech
- Social Science

Ready Reference
- Almanacs
- Calendars
- Dictionaries
- More...

Reading Room
- Books
- Magazines
- Newspapers

KidSpace

TeenSpace

Special Collections
- Literary Criticism
- Blogs
- More...

Searching Tools

About the IPL

Ask a Question

Antiques and Collectibles Appraisal

So, you've inherited a wonderful whats-it from your great aunt Verbena. Or perhaps you bought a terrific thing-a-ma-jig at a garage sale. But what is it, and more importantly, what is it *worth*? Read on, and find the way to begin your research in the following pathfinder.

Places to Visit on the Internet

Antiques Roadshow Online
The popular TV show goes online. There are many useful sections. One is "Tips of the Trade," which details a particular facet of collecting each month, and the previous articles are still accessible. Another section, "Antique Speak," is a small, but growing, glossary of antique terms. Make sure to check out "Stories from the Road," where you can get sneak previews of appraisals on upcoming shows.
http://www.pbs.org/wgbh/pages/roadshow/

Ebay
Ebay is an excellent resources for beginning collectors. Often, sellers have very good pictures in their ads, which can help with identification of your item. Also, for *up to the minute* pricing of items Ebay is *the* place to go. Some people have not appreciated the advent of online auctioning of antiques. Certainly, fakes and reproductions are much easier to pass online than in person. However, if you want to know what your item is currently selling for, and what *you* might reasonably hope to get for it, then look on ebay.
http://www.ebay.com/

Kaleden
You'll find several useful features on this site. Antique dealers/members of the site put items up for auction. You don't need to buy anything, but cruise through the different galleries and look at the incredible pictures, which you can enlarge to full-screen view. Another not to miss section is entitled "Online Publications" which features articles from 19 antiques publications. Make sure to browse through "Antiques & Collecting Magazine."
http://www.kaleden.com/

Sotheby's This is the site for one of the world's premiere auction houses. If you have what you believe to be a really high ticket item, take advantage of their appraisal service.
http://search.sothebys.com/services/ for appraisal.
For main site: http://www.sothebys.com/

Figure 3-5. Exemplary Pathfinders, Internet Public Library

Internet Public Library
Antiques
http://ipl.si.umich.edu/div/pf/entry/48439/

1. A search engine is available to visitors so they may search the more than 100 pathfinders available.
2. Users who prefer browsing can use the breadcumb trail employed on this Web site (Home > Pathfinders).
3. All Web addresses are typed out so that this pathfinder can be usable as a printed document.
4. Annotations are consistent in length.
5. The introduction is very practically written.

Figure 3-6. Exemplary Pathfinders, Las Vegas-Clark County (NV) Library District

Las Vegas-Clark County Library District
Science Projects
www.lvccld.org/ref_info/info_guides/science_projects.htm

1. Bold colors and splashy background were designed especially for kids.
2. The use of cover art and images depicting themes requires less reading for visitors.
3. The library's logo fits nicely into this design.
4. The use of popular tabs makes navigating easier for kids.
5. Each title is linked to the library catalog.

Pathfinders

A B C D E F G H I J K L M N O P Q R S T U V W X Y Z

Civil War

Circulating Materials | Reference Sources
Electronic Sources | Web Sites

Click HERE for a printable copy of this Pathfinder.

Circulating Materials

- Civil War A toZ - **Call Number 973.7/Bol**
This book contains short entries on important people, places, issues and events of the Civil War.

- Gettysburg - **Call number 973.7/Civ**
This book is one of a large set that covers a wide variety of topics. Each book has many pictures and maps in addition to the text. There is also an extensive index to the series.

- The Photographic History of the Civil War - **Call number 973.7/Pho**
If you need pictures, this old multi-volume set is a gold mine.

- Songs and Stories of the Civil War - **Call number 973.7/Sil**
This book tells the story of the war through the music of the time and includes lyrics and sheet music for each song.

- Civil War Generals of the Confederacy - **Call number 973.7092/Reg**
Robert E. Lee, Stonewall Jackson, James Longstreet, J. E. B. Stuart, and Nathan Bedford Forrest are covered in this volume.

Figure 3-7. Exemplary Pathfinders, Massillon (OH) Public Library

Massillon Public Library
Civil War
www.massillon.lib.oh.us/ya/homework/civil_war.htm

1. An alphabetical index is available for easy navigation.
2. Graphics are appropriate for the grade level of the intended audience.
3. Printer-friendly pages are made available in PDF format.
4. Links are made available to other pages suited to the intended audience.
5. Bibliographic units are categorized by format and accessibility.

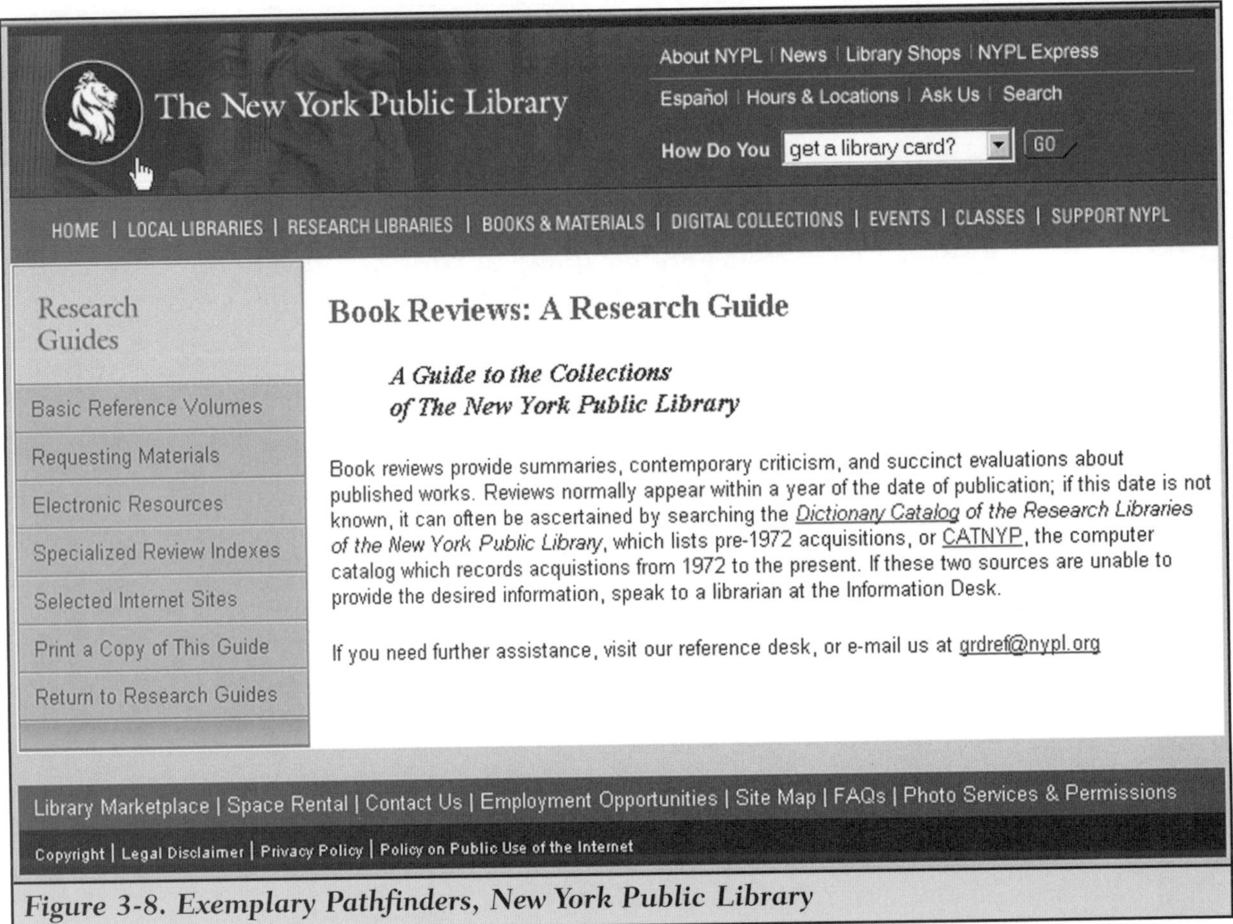

Figure 3-8. Exemplary Pathfinders, New York Public Library

New York Public Library
Book Reviews: A Research Guide
www.nypl.org/research/chss/grd/resguides/bkrev/

1. The homepage of the pathfinder fits on one screen and requires no scrolling.
2. An introductory paragraph includes scope of collection.
3. Each item in the table of contents is a unique Web page.
4. A printer-friendly copy of the guide is available with visible URLs.
5. The main navigation bar is available in addition to a link back to all research guides.

Part II

100 Ready-To-Use Pathfinders

Adoption: A Research Guide

Adoption is a proceeding that creates a legal relation between a parent and a child. Adoption is generally done through public and private organizations or through direct placement with the help of a lawyer. The procedure is often complex, as are the many issues that develop around it. Many books and organizations exist to assist adoptive parents or provide support for adopted children. The terms and phrases listed in the subject headings below can be used to search for more materials in the library's catalog and research databases. If you need further assistance, please ask a librarian.

Books for Adults

Adopting on Your Own: The Complete Guide to Adopting as a Single Parent by Lee Varon. New York: Farrar, Straus, & Giroux, 2000. ISBN: 0374128839.
Offers practical advice for single parents who are considering adoption.

The Adoption Resource Book, 4th ed. by Lois Gilman. New York: Harper Perennial, 1998. ISBN: 0062733613.
Information on financing, international adoption, and a listing of more than 1,000 agencies and support groups.

The Encyclopedia of Adoption by Christine A. Adamec and William L. Pierce. New York: Facts On File, 2000. ISBN: 0816040419.
This A-to-Z reference addresses the social, legal, economic, psychological, and political aspects of the adoption experience.

How to Adopt Internationally: A Guide for Agency-Directed and Independent Adoptions by Jean Nelson Erichsen and Heino R. Erichsen. Fort Worth, TX: Mesa House, 2003. ISBN: 094035215X.
Leads readers through every phase of the international adoption process from finding an agency and organizing a home study to choosing a country.

Is Adoption for You?: The Information You Need to Make the Right Choice by Christine A. Adamec. New York: J. Wiley, 1998. ISBN: 0471183121.
Provides prospective adoptive parents with the information and inspiration for deciding whether to adopt.

Telling the Truth to Your Adopted or Foster Child: Making Sense of the Past by Betsy Keefer and Jayne E. Schooler. Westport, CT: Bergin & Garvey, 2000. ISBN: 0897896912.
Offers background and practical information on discussing an adoptive child's history.

There Are Babies to Adopt: A Resource Guide for Prospective Parents by Christine A. Adamec. New York: Citadel Press, 2002. ISBN: 0806523344.
Instructional manual for married couples or singles wishing to adopt infants.

Books for Children and Young Adults

Allison by Allen Say. Boston: Houghton Mifflin, 1997. ISBN: 039585895X.
Allison realizes that she looks more like her favorite doll than like her parents. She comes to terms with this discovery through the help of a stray cat.

Find a Stranger, Say Goodbye by Lois Lowry. New York: Laurel-Leaf, reprint ed., 1990, 1978. ISBN: 0440205417.
A teenage girl deals with not knowing the identity of her biological mother.

Heaven by Angela Johnson. New York: Simon Pulse, 2000. ISBN: 0689822901.
Marley, 14, finds out that the parents whom she has known and loved are not hers by birth.

I Love You Like Crazy Cakes by Rose A. Lewis. Boston: Little, Brown, 2000. ISBN: 0316525383.
A woman describes how she went to China to adopt a baby.

Over the Moon: An Adoption Tale by Karen Katz. New York: Henry Holt, 1997. ISBN: 0805050132.
A long-awaited baby is born, and the adoptive parents who have been dreaming of her fly far away to bring her home.

Tell Me Again About the Night I was Born by Jamie Lee Curtis and Laura Cornell. New York: HarperCollins, 1996. ISBN: 006024528X.
A young girl asks her parents to tell her again about the cherished family story of her birth and adoption.

Magazine Articles

"Blessed by adoption: More Americans are adopting than ever before. Their stories are heartwarming sagas of hope, faith, and love," by Jennifer Wilson. *Better Homes and Gardens*, February, 2003, v81, i2, p122.

"The long road home (adopting children abroad)," by Catherine Siskos. *Kiplinger's Personal Finance Magazine*, December 2000, v54, i12, p82.

"Priceless: That's how adoptive parents describe their children. But adoption is also a financial transaction. A look at the intersection of money and miracles," by Gay Jervey. *Money*, April 1, 2003, v32, i4, p119.

Web Sites

About.com: Adoption
http://adoption.about.com
Includes support services and forums for adoption professionals and parents. Also contains a directory of legal information arranged by state.

Adoption.com
www.adoption.com
Contains information for parents, women who are pregnant, and those seeking reunions. Also includes directories of adoption professionals and discussion groups.

National Adoption Center
www.adopt.org
Supports all aspects of adoption, particularly for children with special needs and those from minority cultures.

National Adoption Information Clearinghouse
http://naic.acf.hhs.gov
Sponsored by the Administration for Children & Families, U.S. Department of Health & Human Services, this site includes a national directory of agencies, services, state officials, and support groups.

Subject Headings

- adoption
- adoption—law and legislation
- children, adopted
- intercountry adoption
- intercountry adoption—law and legislation
- interracial adoption

Advertising: A Research Guide

The resources listed below will assist researchers in finding advertisement information including primary documents such as historic ads, agency contacts, and ratings. This guide also serves to acquaint the researcher with some of the basic informational sources on the topic. The terms and phrases listed in the subject headings below can be used to search for more materials in the library's catalog and research databases. If you need further assistance, please ask a librarian.

Reference Materials

*Ad*Access*. John W. Hartman Center for Sales, Advertising, and Marketing History, Duke University Press. 1999.
http://scriptorium.lib.duke.edu/adaccess/
Features images from over 7,000 advertisements printed in U.S. and Canadian newspapers and magazines between 1911 and 1955 on topics of radio, television, transportation, beauty and hygiene, and World War II.

Advertising Age Encyclopedia of Advertising edited by John McDonough and Karen Egolf. 3 vols. New York: Fitzroy Dearborn, 2002. ISBN: 1579581722.
Contains surveys of agencies and advertisers, past and contemporary, as well as their ad campaigns. Also includes biographies, issues, methodology, and strategy.

The Emergence of Advertising in America: 1850–1920. John W. Hartman Center for Sales, Advertising, and Marketing History. Duke University Press. 2000.
http://scriptorium.lib.duke.edu/eaa/browse.html
Presents over 9,000 images relating to the early history of advertising in the United States. The materials are drawn from the Rare Book, Manuscript, and Special Collections Library at Duke University.

Directories

Advertising Age's FactPack
www.AdAge.com, QwikFIND: aan86x (PDF)
This annual directory lists marketing industry data such as who spends the most on advertising, what is the cost of a TV ad, total amount spent on advertising in different media, and more.

The Advertising Red Books. Advertiser and Advertising Agencies. (Also international edition.) New Providence, NJ: Lexis-Nexis, 2003– . Semiannual.
The Agency Directory includes profiles of nearly 14,000 U.S. and international advertising agencies, including major accounts represented by each agency, fields of specialization, breakdown of gross billings by media, and contact information on agency personnel. *The Advertiser Directory* contains information on over 15,000 U.S. and international advertisers who each spend more than $200,000 annually on advertising.

Standard Rate and Data Service. Des Plaines, IL: SRDS. Serial publications.
A series of directories with varying titles that include media rates and information including business, consumer publications and newspapers, ad rates and specifications; broadcast stations, their formats and audience demographics, direct marketing lists, and more.

Journals

Advertising Age. Chicago: Crain Communications, 1930– . Weekly. ISSN: 00018899.

Journal of Advertising. Athens, GA: American Academy of Advertising, 1960– . Quarterly. ISSN: 00913367.

Journal of Advertising Research. Advertising Research Foundation. Nyack, NY: Cambridge University Press, 1960– . Quarterly. ISSN: 00218499.

Journal of Marketing. New York: American Marketing Association, 1936– . Quarterly. ISSN: 00222429.

Journal of Marketing Research. Chicago, American Marketing Association, 1964– . Quarterly. ISSN: 00222437.

Web Sites

AC Nielsen
www2.acnielsen.com/site/index.shtml
Provides a variety of free and fee-based information on consumers, markets, and product information.

Nielsen//NetRatings
www.nielsen-netratings.com
Provides information on Web advertising and online usage behaviors.

Organizations

The Advertising Council
www.adcouncil.org
Leading producer of public service advertisements (PSAs) since 1942. Addresses social issues.

The Advertising Research Foundation
www.arfsite.org
Founded in 1936, The Advertising Research Foundation is a nonprofit corporate-membership association that represents more than 400 advertisers, advertising agencies, research firms, media companies, educational institutions and international organizations.

American Advertising Federation
www.aaf.org
Trade association that represents 50,000 professionals in the advertising industry.

American Association of Advertising Agencies (AAAA)
www.aaaa.org
Founded in 1917, the AAAA is the national trade association and represents large, multinational agencies. Membership produces approximately 75% of the total advertising volume placed by agencies nationwide.

Subject Headings

- advertising—social aspects—United States
- advertising—statistics
- advertising—United States—history
- advertising agencies—directories
- advertising agencies—statistics
- advertising media planning—Africa—directories
- radio advertising—United States—directories
- slogans
- television advertising—United States—directories

African-American History: A Research Guide

Learn about the importance of the Dred Scott case, the Civil War, the civil rights movement, school integration, and how these past events affect the present-day African-American community by exploring the booklists, online databases, Web sites and other resources assembled by library staff. The terms and phrases listed in the subject headings below can be used to search for more materials in the library's catalog and research databases. If you need further assistance, please ask a librarian.

Reference Books

African American Literary Criticism 1773–2000 by Hazel Arnett Ervin. New York: Twayne, 1999. ISBN: 0805716831.
Includes major periods and themes of African-American literary criticism through more than 100 essays and articles. Also includes a chronology.

Africana: The Encyclopedia of the African and African American Experience edited by Kwame Anthony Appiah and Henry Louis Gates. New York: Basic Civitas Books, 1999. ISBN: 0465000711.
Landmark reference includes illustrations, charts, graphs, pictures, and articles covering the religion, arts, and cultural life of Africans and of black people everywhere.

Encyclopedia of African and African-American Religion by Stephen D. Glazier New York: Routledge, 2001. ISBN: 0415922453.
Covers religious movements and churches of Africa, North America, South America, and the Caribbean.

Encyclopedia of African-American Culture and History edited by Jack Salzman, David Lionel Smith, and Cornel West. New York: Macmillan Library Reference, 1995. ISBN: 0028973453; Supplement, 2000. ISBN: 0028654412.
An excellent starting point for research. Includes biographies of African-Americans and essays on historical periods, places, cultural achievements, professions, and sports. The supplement includes updates to articles and more than 100 new entries.

The Harvard Guide to African American History edited by Evelyn Brooks Higginbotham, Leon F. Litwack, and Darlene Clark Hine. Cambridge, MA: Harvard University Press, 2001. ISBN: 0674002768.
Includes essays on historical research aids, comprehensive bibliographies, and directories of library collections. Also features Web sites, photo archives, and film repositories.

The Oxford Companion to African American Literature edited by William L. Andrews, Frances Smith Foster, and Trudier Harris. New York: Oxford University Press, 1997. ISBN: 0195065107.
Contains entries on African-American authors, literary works, and themes or topics addressed in African-American literature.

Web Sites

African-American Inventors Database
www.detroit.lib.mi.us/glptc/aaid/index.asp
Published by the Detroit Public Library and searchable by invention, inventor or patent number.

The African-American Mosaic
www.loc.gov/exhibits/african/intro.html
Published by the Library of Congress, this Web site chronicles African-Americans in history through colonization, abolition, migrations, and the WPA.

The African Presence in the Americas, 1492–1992.
www.si.umich.edu/CHICO/Schomburg/
An exhibition portfolio from the Schomburg Center for Research in Black Culture, introduces researchers to the dynamics and dimensions of the 500-year history of African people in the Americas.

Black History Month
www.galegroup.com/free_resources/bhm/index.htm
Provides free resources for teachers and student including biographies, selected works of literature and their significance, quiz, activities and time line.

Finding Records of Your African American Ancestors 1870 to Present
www.familysearch.org/Eng/Search/RG/guide/36367_AfricanAmer2.asp
A guide to finding your ancestors published by the Church of Jesus Christ of Latter-day Saints.

Harlem, 1900–1940: An African-American Community
www.si.umich.edu/CHICO/Harlem/index.html
An exhibition portfolio from the Schomburg Center for Research in Black Culture, this exhibit offers activism, arts, business, community, sports, intellectuals, and writers.

Schomburg Center for Research in Black Culture
www.nypl.org/research/sc/sc.html
The Schomburg Library is devoted to the preservation of African-American photographs and text.

Subject Headings

- African American education
- African American laborers
- African American music
- African American organizations
- African American publishers
- African American railroad employees
- African American soldiers
- African American translators
- African American writers
- African Americans—autobiography
- African Americans—relations with Quakers
- African Americans—religious education
- African Americans—satire

American Folklife: A Research Guide

American folklife is the traditional, expressive, shared culture of various groups in the United States. These groups can be grouped by ethnicity, occupation, religious affiliation, and geographic locations. Expressive culture includes a wide range of creative and symbolic forms, such as customs, beliefs, technical skills, languages spoken, drama, ritual, architecture, music, play, dance, drama, ritual, pageantry, and handicraft. These forms of expression can be maintained or perpetuated without formal instruction or institutional direction, as defined by the Library of Congress. This guide introduces the researcher to some of the basic informational sources on the topic. The terms and phrases listed in the subject headings below can be used to search for more materials in the library's catalog and research databases. If you need further assistance, please ask a librarian.

Books

Folklife and Fieldwork: A Laymen's Introduction to Field Work, rev. ed. by Peter Bartis. Washington, DC: Library of Congress, 2002.
Instructional guide on collecting, presenting, and preserving materials that represent local folklife.

Folklore: An Encyclopedia of Beliefs, Customs, Tales, Music, and Art by Thomas A. Green. Santa Barbara, CA: ABC-CLIO, 1997. ISBN: 087436986X.
Includes over 240 in-depth articles that cover historical and contemporary form, figures, and fields of folklore.

Folklore in America by Tristram Potter Coffin and Hennig Cohen. Garden City, NY, Doubleday, 1966.
Includes tales, songs, superstitions, proverbs, riddles, games, folk drama, and folk festivals.

Great American Folklore: Legends, Tales, Ballads, and Superstitions from All Across America by Kemp P. Battle. Garden City, NY: Doubleday, 1986. ISBN: 0385185553.

A Treasury of American Folklore: Stories, Ballads, and Traditions of the People by Benjamin Albert Botkin. New York: Crown, 1944. ASIN: 9992604328.

Web Sites

The American Folklife Center, The Library of Congress
www.loc.gov/folklife/
Clearinghouse for folklife services, information, and guides for the fifty states. Also includes selected collections from the "Archive of Folk Culture" and online collections such as "Voices from the Days of Slavery: Former Slaves Tell Their Stories" and "Folk-Songs of America: The Robert Winslow Gordon Collection, 1922–1932."

Archives of Traditional Music
www.indiana.edu/~libarchm/index.html
Includes commercial and field recordings of vocal and instrumental music, folktales, interviews, oral history, videotapes, photographs and manuscripts.

The Center for Folklife Programs and Cultural Studies
www.folkways.si.edu
Contains the Moses and Frances Asch Archives (recordings, business records, correspondence, etc. from the Folkways Records) and the documentation done by scholars for Smithsonian projects, exhibits, and the annual Festival of American Folklife held on the Mall in Washington, DC.

Tapnet Links (National Endowment for the Arts, National Council for the Traditional Arts)
http://afsnet.org/tapnet
Includes links to folk arts in education, museums and archives, federal and national folk and traditional arts programs, and state and regional folk arts programs and links.

A Teacher's Guide to Folklife Resources
www.loc.gov/folklife/teachers/
Information on many useful print and electronic publications for folklife educators, including contact and ordering information.

Journals and Magazines

American Quarterly. Baltimore: Johns Hopkins University Press, 1949– . Quarterly. ISSN: 00030678.
Publishes essays that examine American societies and cultures, past and present, in global and local contexts.

Folklore Forum. Bloomington, IN: Folklore Forum Society, 1968– . Semiannually. ISSN: 00155926.
A communication for students of folklore.

Journal of American Folklore. New York: Published for the American Folklore Society, 1888– . Quarterly. ISSN: 00218715.
Publishes scholarly articles, essays, notes, and commentaries, reviews of books, exhibitions and events, sound recordings, film and videotapes.

Journal of Folklore Research. Bloomington, IN: Indiana University Folklore Institute, 1983– . Three times a year. ISSN: 07377037.
Devoted to the study of the world's traditional creative and expressive forms, the Journal of Folklore Research provides an international forum for current theory and research among scholars of folklore and related fields.

Journal of Popular Culture. Bowling Green, OH: Bowling Green State University Press, 1967– . Quarterly. ISSN: 00223840.
The official publication of the Popular Culture Association.

Western Folklore. Los Angeles: California Folklore Society, 1947– . Quarterly. ISSN: 0043373X.
Devoted to the description and analysis of regional, national, and international folklore and custom.

Winterthur Portfolio. Chicago: University of Chicago Press, 1964– . Quarterly. ISSN: 00840416. Also available at *www.journals.uchicago.edu/WP/journal/.*
Interdisciplinary journal committed to fostering knowledge of the American past by publishing articles on the arts in America and the historical context within which they developed.

Subject Headings

- festivals
- folklore—United States
- legends—United States
- manners and customs
- rites and ceremonies
- tales—United States
- Add "juvenile literature" to limit search to children's materials.

American History: A Research Guide

Researchers looking for information from the Crittenden Compromise, the Hartford Convention, and the Sons of Liberty to the Declaration of Independence can find a vast amount of information in the resources listed here. The terms and phrases listed in the subject headings below can be used to search for more materials in the library's catalog and research databases. If you need further assistance, please ask a librarian.

Primary Source Material

Core Documents of U.S. Democracy
www.gpoaccess.gov/coredocs.html
Includes full-text documents like the Articles of Confederation, Bill of Rights, Emancipation Proclamation, and the Gettysburg Address. Also includes documents related to presidencies and legislative and judicial history.

Great American Speeches: 80 Years of Political Oratory
www.pbs.org/greatspeeches
An archive of speeches and background information from 1900–1999.

Library of Congress American Memory Project
http://lcweb2.loc.gov/ammem
Primary source materials relating to the history and culture of the United States.

Dictionaries and Encyclopedias

American Eras edited by Gretchen D. Starr-LeBeau, Jessica Kross, and Robert J. Allison. 8 vols. Detroit: Gale Research, 1997– . ISBN: 0787614777.
Includes essays on events, publications, lifestyles, and individuals important to American history. Vol. 1: Early American civilizations and exploration to 1600; vol. 2: The colonial era, 1600–1754; vol. 3: The Revolutionary era, 1754–1783; vol. 4: Development of a nation, 1783–1815; vol. 5: The reform era and eastern U.S. development, 1815–1850; vol. 6: Westward expansion, 1800–1860; vol. 7: Civil War and Reconstruction, 1850–1877; vol. 8: Development of the industrial United States, 1878–1899.

Annals of America. Chicago: Encylopaedia Britannica, 1968–1987. 20 vols.
A multivolume encyclopedia set that includes the words of more than 1,500 authors including illustrations, photographs, articles, essays, speeches, songs, excerpts from books, and primary source documents.

Datapedia of the United States. 1790–2005: America Year by Year, 3rd ed. Lanham, MD: Bernan Press, 2004. ISBN: 0890598622.
Represents data from several hundred indicators of social, economic, political, and cultural developments. Consists chiefly of tables with bibliographical references and index.

Dictionary of American History, 3rd ed. edited by Stanley I. Kutler. 10 vols. New York: Charles Scribner's Sons, 2002. ISBN: 0684805332.
Alphabetical entries with coverage extending from the colonial period to the present including maps and illustrations. First published in 1940 and completely revised in this edition.

Encyclopedia of the American Civil War: A Political, Social, and Military History edited by David S. Heidler and Jeanne T. Heidler. 5 vols. Santa Barbara, CA: ABC-CLIO, 2000. ISBN: 1576070662.
Coverage of broad areas of specific topics on the Civil War including photographs, lithographs, and primary source documents.

Encyclopedia of American Cultural and Intellectual History edited by Mary K. Cayton and Peter W. Williams. 3 vols. New York: Charles Scribner's Sons, 2001. ISBN: 0684805618.

Covers historical periods from 1771–2000, including subjects which form the country's rich history such as people, places, politics, arts, economy, social order, the arts, religion, and nature.

Encyclopedia of American History edited by Gary B. Nash. 11 vols. New York: Facts On File, 2003. ISBN: 081604371X.

Details the people, events, and ideas that have shaped the nation. Organized in chronological order.

Encyclopedia of American Social History edited by Mary K. Cayton, Elliot J. Gorn, and Peter W. Williams. 3 vols. New York: Charles Scribner's Sons, 1993. ISBN: 0684192462.

180 essays highlighting the people of the American social structure through ethnology, gender study, geography, literature, religion, anthropology, and sociology.

Encyclopedia of the North American Colonies edited by Jacob E. Cooke. 3 vols. New York: Charles Scribner's Sons, 1993. ISBN: 0684192691.

Covers geography and ecology, patterns of settlement, law and government, economy, society, religion, popular culture, and family life.

Encyclopedia of the United States in the Nineteenth Century edited by Paul Finkelman. 3 vols. New York: Charles Scribner's Sons, 2001. ISBN: 0684805006.

600 articles written by scholars including illustrations, maps, and time lines. Written to support the National Standards for U.S. History.

Encyclopedia of Urban America: The Cities and Suburb edited by Neil Larry Shumsky. 2 vols. Santa Barbara, CA: ABC-CLIO, 1998. ISBN: 0874368464.

Provides a comprehensive view of American cities in a series of 547 signed articles including biographical articles and topical entries.

Web Sites

History Channel
www.historychannel.com
Look for "America's Timeline" and choose a decade.

Making of America
www.hti.umich.edu/m/moagrp/
This Web site, created by the University of Michigan, includes primary source material from the antebellum period through Reconstruction.

Our Documents
www.ourdocuments.gov
Web site promoting the 100 milestone documents of American history.

Smithsonian National Museum of American History
http://americanhistory.si.edu
Includes Information on collections and exhibits.

Subject Headings

- United States—civilization
- United States—history
- United States—intellectual life
- United States—politics and government
- United States—social conditions
- United States and Civil War
- United States and Industrial Revolution
- United States and women and history

Ancient Cultures: A Research Guide

As agriculture further developed around 9000 B.C., people began to settle in one place, since they were not as nomadic and could grow their own food. These settlements began the world's first civilizations. Many materials exist to help us better understand the everyday life of ancient of civilizations and the events that shaped the people and places of this time. The materials below will help researchers to begin their research. The terms and phrases listed in the subject headings below can be used to search for more materials in the library's catalog and research databases. If you need further assistance, please ask a librarian.

Reference Books

The Cambridge Ancient History, 3rd ed. edited by Iorwerth Eiddon Stephen Edwards. 14 vols. New York: Cambridge University Press, 1970.
Originally published between 1924 and 1939, this definitive work covers all time periods from prehistory to late antiquity.

Civilization of the Ancient Mediterranean: Greece and Rome by Michael Grant and Rachel Kitzinger. 3 vols. New York: Charles Scribner's Sons, 1988. ISBN: 0684175940.
Entries arranged by broad topics like history, population, agriculture, technology, government, and society.

Chronologies and Time Lines

Chronology of the Ancient World, 2nd ed. by Elias Joseph Bickerman. Ithaca, NY: Cornell University Press, 1980. ISBN: 080141282X.

Great Events from History: Ancient and Medieval edited by Frank Northen Magill. Series (3 vols). Englewood Cliffs, NJ: Salem Press, 1972.
Covers these time periods: vol. 1. 4000–1 B.C.; vol. 2. A.D. 1–950.; vol. 3. 951–1500.

Smithsonian Timelines of the Ancient World edited by Christopher Scarre Washington, DC: Smithsonian Institution, 1993. ISBN: 1564583058.
Chronology of the ancient world spanning the first life on earth, the ice age, crossing to America, the age of monuments, imperial realms, and conquests and crusades.

Specialized Encyclopedias and Dictionaries

Biographical Dictionary of Ancient Greek and Roman Women: Notable Women from Sappho to Helena by Marjorie and Benjamin Lightman. New York: Facts On File, 2000. ISBN: 0816031126.
Over 400 entries illustrate the lives of women including poets, wives, and mistresses to rulers, slaves, and businesswomen from archaic Greece in the 6th century BCE to the fall of Rome in 476 CE.

Encyclopedia of Ancient Civilizations by Arthur Cotterell. New York: Mayflower Books, 1980. ISBN: 083172790X.
Covers the ancient civilizations of Egypt, West Asia, India, Europe, China, and America.

Encyclopedia of the Roman Empire, rev. ed. edited by Matthew Bunson. New York: Facts On File, 2002. ISBN: 0816045623. Covers the personalities, places, events, and ideas of this time period.

Larousse Encyclopedia of Ancient and Medieval History edited by Marcel Dunan. New York: Excalibur Books, 1981, 1964. ISBN: 0896730832.
An illustrated overview covering ancient times to the 15th century in Europe and the rise of the Aztec and Incan empires in America.

Oxford Classical Dictionary, 3rd ed. rev. edited by Simon Hornblower and Antony Spawforth. New York: Oxford University Press, 2003. ISBN: 0198606419.
Six thousand entries ranging from long articles to brief identifications, covering topics like bee-keeping, botany, magic, Roman law, religious rites, postal service, slavery, navigation, and the reckoning of time.

Oxford Encyclopedia of Ancient Egypt edited by Donald B. Redford. 3 vols. New York: Oxford University Press, 2001. ISBN: 0195102347.
Covers all aspects of Egyptology including 600 articles on history, archaeology, economics, science, and literary and religious studies.

Web Sites

BBC Online Ancient History
www.bbc.co.uk/history/ancient
Explore ancient history through feature articles, 3-D virtual tours, games and animations.

PBS on the Ancient World
www.pbs.org/history/history_ancient.html
A PBS series dedicated to ancient civilizations and topics such as the ice cities of the Inca and Peter and Paul and the Christian revolution.

The Seven Wonders of the Ancient World
www.infoplease.com/ipa/A0001327.html
Provides a listing of the seven wonders of the ancient world.

Subject Headings

- names of country of antiquity (i.e. Rome—Antiquities)
- names of ancient races and peoples (i.e. Hittites)
- names of individuals (i.e. Tutankhamen)
- ancient Greece (or Rome, or Egypt, etc.)
- civilization—history
- civilization, ancient
- civilization, classical
- costume—history
- Greece (or Rome or Egypt, etc.), antiquities
- history, ancient

Ancient Egypt: A Research Guide

Ancient Egypt was the birthplace of one of the world's first civilizations that included a thriving culture, which contributed the world's first national government, basic forms of arithmetic, and a 365-day calendar. The Egyptians invented a form of picture writing called hieroglyphics and papyrus and developed one of the first religions to emphasize life after death. The terms and phrases listed in the subject headings below can be used to search for more materials in the library's catalog and research databases. If you need further assistance, please ask a librarian.

Reference Books

Cultural Atlas of Ancient Egypt edited by John Baines and Jaromír Málek. New York: Facts On File, 2000. ISBN: 0816040362.
Introduction to life in ancient Egypt.

Egyptology: An Introduction to the History, Art and Culture of Ancient Egypt edited by James Putnam. New York: Crescent Books, 1996. ISBN: 185627845X.
Covers interesting points about ancient Egypt, with a special emphasis on life and art.

Handbook to Life in Ancient Egypt edited by A. Rosalie David. New York: Facts On File, 1998. ISBN: 0816033129.
Includes historical background, geography, society and government, religion of the living, funerary beliefs and customs, architecture and building, written evidence, the army and navy, foreign trade and transport, economy and industry, and everyday life.

Life in Ancient Egypt edited by Thomas Streissguth. San Diego: Lucent Books, 2001. ISBN: 1560066431.
Provides informative details on how the Egyptians lived and what they believed.

The Oxford Encyclopedia of Ancient Egypt by Donald B. Redford. 3 vols. New York: Oxford University Press, 2001. ISBN: 0195102347.
Covers all aspects of Egyptology including 600 articles on history, archaeology, economics, science, and literary and religious studies.

The Oxford History of Ancient Egypt edited by Ian Shaw. New York: Oxford University Press, 2000. ISBN: 0192802933.
Provides detailed historical coverage of Egypt from the early Stone Age to its incorporation into the Roman Empire. Also examines cultural and social patterns, including stylistic developments in art and literature.

What Life Was Like on the Banks of the Nile: Egypt, 3050–30 B.C. Alexandria, VA: Time-Life Books, 1996. ISBN: 0809493780.
Tells of the day to day activities of people in ancient Egypt.

Mythology and Religion

The Gods of Ancient Egypt edited by Barbara Watterson. Stroud: Sutton, 1999. ISBN: 0750922257.
Explores the main characteristics of ancient Egyptian religions and traces the origins of all the major deities of ancient Egypt.

Handbook of Egyptian Mythology edited by Geraldine Pinch. Santa Barbara, CA: ABC-CLIO, 2002. ISBN: 1576072428.
Survey of Egyptian mythology spanning ancient Egyptian culture (circa 3200 B.C. to A.D. 400).

Pyramids

Pyramids by Stuart A. Kallen. San Diego: Lucent Books, 2002. ISBN: 156006773X.
Gives brief information on how the pyramids were built with illustrations and pictures to supplement the basic data.

The Pyramids of Giza by Tim McNeese. San Diego: Lucent Books, 1997. ISBN: 1560064269.

Describes the construction of the three great pyramids on the Giza Plateau near the edge of Egypt's Western Desert.

Web Sites

Ancient Egypt
www.ancientegypt.co.uk/menu.html

Created by the British Museum, this Web site provides general information on many subtopics like mummification, pharaohs, hieroglyphics, pyramids, and temples.

Mummies of Ancient Egypt
www.si.umich.edu/CHICO/mummy/

Answers questions like "what are mummies?" and "how are mummies made?" Also contains information about the ancient Egyptians' belief in the afterlife.

Pyramids: The Inside Story
www.pbs.org/wgbh/nova/pyramid/

Companion Web site to the PBS *NOVA* television series about exploring and excavating pyramids.

The Religion of Ancient Egypt
www.mnsu.edu/emuseum/prehistory/egypt/religion/religion.html

Includes detailed information on specific deities, priests' theology, and a list of gods in the astrology of ancient Egypt.

Subject Headings

- building—Egypt
- Egypt—antiquities
- Egypt—civilization—332 B.C.–638 A.D.
- Egypt—civilization—to 332 B.C.
- Egypt—history—to 332 B.C.
- Egypt—social life and customs—to 332 B.C.
- Egyptology
- pyramids—Egypt
- pyramids of Giza (Egypt)—design and construction

Animals: A Research Guide

Animals come in all shapes and sizes and live in many different habitats. The resources listed below will help you find information about where animals live, how they reproduce and take care of their young, and their migration and interaction with other animals and humans. The terms and phrases listed in the subject headings below can be used to search for more materials in the library's catalog and research databases. If you need further assistance, please ask a librarian.

Reference Books

Animal: Definitive Visual Guide to the World's Wildlife edited by Don E. Wilson and David Burnie. New York and Washington, DC: Dorling Kindersley, Smithsonian Institution, 2001. ISBN: 0789477645.
Includes profiles and images from top 70 biologists of over 2,000 species represented in the book.

Encyclopedia of the World's Zoos edited by Catharine E. Bell. 3 vols. Chicago: Fitzroy Dearborn, 2001. ISBN: 1579581749.
History and background of more than 150 of the world's major zoos, types of animals housed, concepts and issues of animal care and keeping, facility design, and professional organizations.

Encyclopedia of Animal Rights and Animal Welfare edited by Marc Berkoff and Carron Meaney. Westport, CT: Greenwood Press, 1998. ISBN: 0313299773.
Essays cover issues, controversies, significant historical figures, and ideologies related to the treatment of animals. Includes a directory of organizations.

Grzimek's Animal Life Encyclopedia, 2nd ed. edited by Bernhard Grzimek. 17 vols. Detroit: Gale Group, 2003. ISBN: 0787653624.
Standard reference work covers both biological and behavioral instincts. Vol. 1: Lower Animals; vol. 2: Insects; vol. 3: Mollusks and Echinoderms; vol. 4: Fishes I; vol. 5: Fishes II/Amphibians; vol. 6: Reptiles; vols. 7–9: Birds I–III; vols. 10–13: Mammals I–IV.

The Illustrated Encyclopedia of Wildlife edited by Mary C. Pearl. 15 vols. Lakeville, CT: Grey Castle Press, 1991. ISBN: 1559050527.
Includes informative reading and the full-color illustrations of wildlife. Arranged by class, with mammals (vols. 1–5), birds (vols. 6–8), reptiles and amphibians (vol. 9), fishes (vol. 10), and invertebrate classes (vols. 11–15).

The Kingfisher Illustrated Nature Encyclopedia by David Burnie. New York: Kingfisher, 2004. ISBN: 0753455765.
An illustrated guide to the world's many groups of plants and animals.

Reader's Digest North American Wildlife edited by Susan J. Wernert. Pleasantville, NY: Reader's Digest Association, 1982. ISBN: 0895771020.

Magazines

National Wildlife. Washington, DC: National Wildlife Federation. 1963– . Bimonthly. ISSN: 00280402.
Covers topics like conservation, natural history, outdoor life and wildlife. Includes dazzling photography.

Ranger Rick. Vienna, VA: National Wildlife Federation. 1983– . Monthly. ISSN: 07386656.
Wildlife magazine for ages 7 and up.

Web Sites

2003 IUCN Red List of Threatened Species
www.redlist.org
Allows searching of comprehensive database of threatened species.

ENature.com
www.eNature.com
Sponsored by the National Wildlife Federation, this Web site includes searchable filed guides that allows you to find wildlife by zip code, regional bird guide, and wildlife sighting journal.

Species in Parks: Flora and Fauna Database
http://ice.ucdavis.edu/nps/
A database of vascular plants and vertebrate animals residing in U.S. national parks.

Associations and Organizations

The American Zoo and Aquarium Association
www.aza.org
Nonprofit organization dedicated to the advancement of zoos and aquariums in the areas of conservation, education, science, and recreation.

National Wildlife Federation
www.nwf.org
Official Web Site of the National Wildlife Federation.

National Zoo
http://nationalzoo.si.edu
Information about the animals who live at the zoo and their habitat.

U.S. Fish and Wildlife Service
www.fws.gov
Official Web Site of the U.S. Fish and Wildlife Service.

Subject Headings

- animals adaptation
- animals anecdotes
- animals habitations
- animals (name of country)
- animals and civilization
- animals as carriers of disease
- zoo animals
- zoology

Anthropology: A Research Guide

Anthropology is the study of the classification and analysis of humans and their society, descriptively, culturally, historically, and physically. It differs from other fields like sociology because of its inclusion of data from nonliterate peoples and archaeology. The terms and phrases listed in the subject headings below can be used to search for more materials in the library's catalog and research databases. If you need further assistance, please ask a librarian.

Literature Guides

Cultural Anthropology: A Guide to Reference and Information Sources by Josephine Z. Kibbee. Englewood, CO: Libraries Unlimited, 1991. ISBN: 0872877396.
Provides citations and annotations on reference sources, including manuals, bibliographies, indexes, databases, literature surveys and reviews, and dissertations.

Introduction to Library Research in Anthropology, 2nd ed. by John M. Weeks. Boulder, CO: Westview Press, 1998. ISBN: 0813374545.
Includes subject bibliographies, bibliographies, dictionaries, atlases, government documents, manuscripts and archives, dissertations, biographies, and more.

Dictionaries and Encyclopedias

Countries and Their Cultures edited by Melvin Ember and Carol R. Ember. New York: Macmillan Reference USA, 2001. ISBN: 0028649508.
Descriptive summaries of 225 countries and their cultures with bibliographies.

The Dictionary of Anthropology edited by Thomas Barfield. Cambridge, MA: Blackwell, 1997. ISBN: 1557862826.
Focuses primarily on topics in cultural and social anthropology. Emphasizes concepts, theories, and approaches within the discipline of anthropology.

Encyclopedia of Anthropology edited by David E. Hunter. NY: Harper & Row, 1976. ISBN: 0060470941.
Contains illustrations and selective references to major works for most topics.

Encyclopedia of Cultural Anthropology edited by David Levinson and Melvin Ember. New York: Henry Holt, 1996. ISBN: 0805028773.
Encyclopedia of key concepts, methods, and topics central to the study of cultural anthropology with bibliographies.

Encyclopedia of Social and Cultural Anthropology edited by Alan Barnard and Jonathan Spencer. New York: Routledge, 2002. ISBN: 0415285585.
Areas covered include kinship, gender and family, marriage and the body. Provides definitions of terms, concepts and biographical entries.

Encyclopedia of World Cultures edited by David Levinson. 10 vols. Boston: G. K. Hall, 1991– . ISBN: 081688840X.
Sponsored by the Human Relations Area Files. Historical, social, political, economic, linguistic, and religious information are provided with bibliographies. Vol. 1: North America; vol. 2: Oceania; vol. 3: South Asia; vol. 4: Europe; vol. 5: East and Southeast Asia; vol. 6: Russia and Eurasia, China; vol. 7: South America; vol. 8: Middle America and the Caribbean; vol. 9: Africa and the Middle East; vol. 10: Indexes.

Indexes and Abstracts

Abstracts in Anthropology. Farmingdale: Baywood, 1970– . Bimonthly.
Abstracts are grouped in four sections: archaeology, ethnology (or cultural anthropology), linguistics, and physical anthropology.

International Bibliography of Social and Cultural Anthropology. Paris: UNESCO, 1955– . Annual.
Devoted primarily to sociocultural anthropology.

Web Sites

Academic Info Anthropology
www.academicinfo.com/anth.html
Directory of online anthropology resources including digital publications, cultural and social anthropology, physical and biological anthropology, archaeology, as well as a listing of anthropology organizations.

Anthro.Net Research Engine
www.anthro.net
A site dedicated to the study of anthropology and archaeology, providing a directory of worthwhile Web sites.

Anthropology Resources on the Internet
www.anthropologie.net
List of Internet resources that are directly or primarily of anthropological relevance.

Anthropology Review Database
http://wings.buffalo.edu/ARD/geninfo.shtml
Database of reviews covering anthropological publications, including books, audiovisual materials, software and multimedia, exhibits, tourist sites, conferences, and online resources.

Journals

American Anthropologist. American Anthropological Association, 1888– . Quarterly. ISSN: 00027294.
Publishes articles, commentaries, and essays on issues of importance to the discipline; includes reviews of books, films, sound recordings, and exhibits.

Annual Review of Anthropology. Annual Reviews. 1972– . Annual. ISSN: 00846570.
Reviews the year's scholarship in anthropology.

Current Anthropology. Wenner-Gren Foundation for Anthropological Research. 1960– . Quarterly. ISSN: 00113204.
Features papers in areas of social, cultural, and physical anthropology, ethnology and ethnohistory, archaeology and prehistory, folklore, and linguistics.

Associations

American Anthropological Association (AAA)
www.aaanet.org
Primary professional society of anthropologists in the United States since founded in 1902. Official site of the AAA contains information about the organization, jobs and careers, meetings, publications, and special interest groups.

The Society for Applied Anthropology
www.sfaa.net
Aspires to promote the integration of anthropological perspectives and methods in solving human problems through their publications, education, and professional development.

Subject Headings

- anthropology
- anthropology—dictionaries
- anthropology—history
- ethnology—dictionaries
- ethnology—encyclopedias
- human evolution
- sociology

Antiques and Collectibles: A Research Guide

Antiques and collectibles come in all forms including dishes, furniture, pottery, books, cookery, dolls and toys. If you are not sure about the value of an item that you currently own or one you want to purchase, check the materials listed below for matching or similar items. You may find out that the old painting you found in the attic is worth more than your car! The terms and phrases listed in the subject headings below can be used to search for more materials in the library's catalog and research databases. If you need further assistance, please ask a librarian.

Books

Antiques Roadshow Primer: The Introductory Guide to Antiques and Collectibles from the Most Watched Series on PBS edited by Carol Prisant. New York: Workman, 1999. ISBN: 076111775X.

Deals with traditional collectibles, such as furniture, glass, pottery, and silver. Includes tips on telling reproductions from the real thing and a basic introduction to the appraisal process.

Antiques Trader's Antiques and Collectibles edited by Kyle Husfloen. Iola, WI: Antique Trader Books, 2000. ISBN: 0873418905.

Guide for collectors of rare vintage items and contemporary collectibles, includes detailed descriptions and photographs.

Antiquing for Dummies by Ron Zoglin and Deborah Shouse. New York: For Dummies, 1990. ISBN: 0764551086.

Contains advice about talking with antique dealers, participating in auctions, and how to locate hidden treasures at garage sales, flea markets, estate sales, auctions, and antique shops.

Kovels' Antiques and Collectibles Price List by Ralph M. and Terry H. Kovel. New York: Crown, 1981– . Annual. ISSN: 07382405.

Yearly price guide. Includes record setting prices from the last year, and predictions about what items will be "hot" in the future.

Kovels' Yellow Pages by R. and T. Kovel. New York: Random House, 2003. ISBN: 0609806246.

Directory of collectors, appraisers, auction houses and clubs with names, addresses, telephone and fax numbers, e-mail, and Internet addresses.

Maloney's Antiques and Collectibles Resource Directory by David J. Maloney. Radnor, PA: Wallace-Homestead, 1995– . Annual. ISSN: 10838449.

Published every 2 years. Provides some information about specific kinds of antiques and collectibles. It includes Web site and e-mail addresses of clubs, experts, and buyers.

Schroeder's Antiques Price Guide by Sharon Huxford and Bob Huxford. Paducah, KY: Collector Books, 1983– . Annual. ISBN: 1574323105.

Popular price guide with hundreds of sharp photos and updated values of over 50,000 items.

Web Sites

Antiques
www.bbc.co.uk/antiques

The British Antiques Roadshow site. Information on how to properly clean and preserve your items, tips on identifying your item, and what to look for when buying an antique.

Antiques Roadshow
www.pbs.org/wgbh/pages/roadshow/
Online version of popular TV show. Offers sneak previews of appraisals on upcoming shows.

Kaleden
www.kaleden.com
Antique dealers and members of the site put items up for auction. Contains articles from 19 antiques publications.

Kovel's online
www.kovels.com
Contains in-depth articles, many of which are alerts about reproductions. Also has twelve of the authors' books online.

Sotheby's
www.sothebys.com
One of the world's premiere auction houses. Offers appraisal service.

Organizations

American Society of Appraisers
www.appraisers.org
Find an expert appraiser in your area. You can search for appraisers by their specialty.

Appraisers Association of America, Inc.
www.appraisersassoc.org
Professional appraisers association, also with an appraiser database, searchable by location and specialization.

Subject Headings

- antiques—catalogs
- antiques—collectors and collecting
- antiques—United States
- collectibles—United States
- collecting

Asian-Pacific American Heritage: A Research Guide

The diversity of the United States gives way to many ethnic groups that have contributed so much to this country. The materials listed below will provide information on the history, contributions, and identify of Asian-Americans including the issues they face today. The terms and phrases listed in the subject headings below can be used to search for more materials in the library's catalog and research databases. If you need further assistance, please ask a librarian.

Books

Asian Americans: An Interpretive History by Sucheng Chan. Boston: Twayne, 1991. ISBN: 0805784268.
Twayne's immigrant heritage of America series. History of Asians in the United States from 1840 to the 1990s.

Dictionary of Asian American History by Hyung-chan Kim. New York: Greenwood Press, 1986. ISBN: 0313237603.
Traces the immigrant experience from more than ten countries in East and Southeast Asia and covers the cultural, social, economic, and political impact of these groups on United States.

Strangers from a Different Shore, rev. ed. by Ronald Takaki. Boston: Little, Brown, 1998. ISBN: 0316831301.
A popular history of Asian-Americans (Chinese, Japanese, Koreans, Vietnamese, Filipinos, and Indians) that covers how they have made their presence felt in America since the early 1800s.

Yellow: Race in America Beyond Black and White by Frank H. Wu. New York: Basic Books, 2002. ISBN: 0465006396.
Wu examines issues affecting Asian-Americans such as affirmative action, globalization, immigration, and other controversial contemporary issues while confronting stereotypes.

Associations

Association for Asian American Studies
www.aaastudies.org
Founded in 1979 for the purpose of advancing the highest professional standard of excellence in teaching and research in the field of Asian-American Studies and to create better understanding of the Chinese, Japanese, Korean, Filipino, Hawai'ian, Southeast Asian, South Asian, Pacific Islander, and other groups.

Chinese Historical Society of America, San Francisco, California
www.chsa.org
The CHSA was established to promote and preserve the contributions of Chinese-Americans through their museum, bookstore and learning center.

Korean American Coalition
www.kacla.org
Community advocacy organization established in 1983. Its mission is to facilitate the Korean-American community's participation in civic, legislative, and community affairs.

National Association of Japan-America Societies
www.us-japan.org
Provides links to many regional Japanese societies in the United States.

The National Federation of Filipino American Associations
www.naffaa.org
Exists to promote the welfare and well-being of all Filipinos and Filipino-Americans throughout the United States.

Organization of Chinese Americans (OCA)
www.ocanatl.org
Strives to advance the cause and foster public awareness of the needs and concerns of Chinese-Americans in the United States.

Web Sites

Angel Island State Park
www.angelisland.org
Sponsored by the Angel Island Association, this Web site offers detailed information about Angel Island, the immigration station off the coast of California that processed thousands of Asians who came to the United States in the early 1900s.

Ask Asia
www.askasia.org
A wealth of information about Asia, Asian-American culture, and history geared to grades K–12.

Becoming American: The Chinese Experience
www.pbs.org/becomingamerican/chineseexperience.html
Filled with facts and information about the Asian-American experience through eyewitness accounts. Features additional organizations, books, films and videos, and a time line.

Origins of APA Heritage Month
www.infoplease.com/spot/asianintro1.html
Discusses origin of the Asian-Pacific American Heritage month.

Museums

Arthur M. Sackler Gallery and the Freer Gallery of Art. The National Museum of Asian Art, Smithsonian Institution, Washington, DC
www.asia.si.edu
Includes art from many Asian countries such as China, Japan, Korea and India, online exhibitions, calendar of events, and teaching guides.

Asian Art Museum Chong-Moon Lee Center for Asian Art and Culture, San Francisco, California
www.asianart.org
Includes nearly 15,000 treasures spanning 6,000 years of history, representing cultures throughout Asia.

Chinese American Museum, Los Angeles, California
www.camla.org
Offers exhibits that illustrates the historical legacy of Chinese-Americans.

Japanese American National Museum, Los Angeles, California
www.janm.org
Collections from Japanese-Americans include exhibits of photographs, papers, art, transcripts, and more.

Korean American Museum
www.kamuseum.org
Seeks to interpret and preserve its history and culture through its collections and programs.

Subject Headings

- Asian Americans—civil rights
- Asian Americans—history
- Asian Americans—race identity
- Asian Americans—social conditions

Automobile Repair: A Research Guide

Automotive repair information can be found in many books and online sources. Whether you are repairing an older vehicle or seeking information on a newer model, the library has many resources that provide wiring diagrams, estimates on the time needed for mechanics to fix the vehicle, and *electrical and vacuum diagrams.* Materials include books on specific automobiles, books that cover several cars but in less detail, and books about specific auto systems. The titles listed below represent popular manuals that have been published, in some cases, as far back as the early 1900s. During this time, many of the titles have changed or merged. The titles listed below represent the most recent editions. The terms and phrases listed in the subject headings below can be used to search for more materials in the library's catalog and research databases. If you need further assistance, please ask a librarian.

Repair Manuals: Books on More than One Automobile (Listed by Publisher)	*Chilton Manuals General Repair Information.* Published by Delmar Learning (Thomson) *Chilton's Auto Repair Manual, 1968–present.* ISSN: 00693634. *Chilton's Import Car Repair Manual, 1972–present.* ISSN: 10442456. *Chilton's SUV Repair Manual* (prior to 1998 included in title below). *Chilton's Truck and Van Repair Manual, 1971–present.* ISSN: 07420315.
Consumer Total Car Care Manuals	Manuals on specific makes and models. Sample title: *Buick REGAL/CENTURY, 1975–87.*
Quick-Reference Manuals	Provides coverage on repair and maintenance, adjustments and diagnostic procedures for specific systems and components. Sample title: *Brake Specifications and Service, 1990–2000.*
Motor Information Systems General Repair Information	*Motor's Auto Repair Manual,* current ed.: 2000–2004, 67th ed. 2 vols. Features domestic passenger cars. Volume 1 covers specifications and service procedures on 2000–2004 DaimlerChrysler Corporation, Ford Motor Company, and General Motors Corporation models available at time of publication. Volume 2 covers dash gauges, speed controls, wiper systems, passive restraints, dash panel service, anti-lock brakes, tire pressure monitoring system, and active suspension systems. *Motor's Factory Shop Manual, Motor Auto Repair Manual.* *Motor Imported Car Repair Manual.* Earlier titles from 1972 and 1973. *Motor Heavy Truck Repair Manual* (includes medium duty trucks), current ed.: 1997–2002, 15th ed. Covers mechanical specifications and service procedures on 1997–2002 medium and heavy duty truck models available at time of publication. *Motor Imported Car Repair Manual,* current ed.: 1997–2002, 22nd ed. Vol. 1: Mechanical Systems Repair Vol. 2: Chassis Electrical Systems. Covers specifications and service procedures on 1997–2002 Acura, Daewoo, Honda, Hyundai, Infiniti, Isuzu, Kia, Lexus, Mazda, Mitsubishi, Nissan, Subaru, Suzuki, and Toyota models available at time of publication. *Motor Imported Engine Performance and Drivability Manual,* current ed.: 2002–2003, Asian, 8th ed. Vol. 1: Acura, Daewoo, Honda, Hyundai, Infiniti, Isuzu, Kia, Lexus and Mazda Vol. 2: Mitsubishi, Nissan, Subaru, Suzuki, and Toyota

Motor Light Truck and Van Repair Manual, current ed.: 1999–2003.
Covers specifications and service procedures on 1999–2003 DaimlerChrysler Corporation, Jeep, Ford Motor Company, and General Motors Corporation models available at time of publication. Previous titles and mergers: *Motor Truck Repair Manual*, *Motor Light Truck and Van Repair Manual*, and *Motor Truck and Diesel Repair Manual*.

Mitchell Repair Manuals Professional Repair Information

General domestic and import manuals with separate books for air-conditioning and transmission repair. A degree of knowledge is assumed.

Chassis: Domestic Cars, Light Trucks and Vans

Chassis: Import Cars, Light Trucks and Vans

Electrical Component Locator: Domestic Cars, Light Trucks and Vans

Engine Performance: Domestic Cars, Light Trucks and Vans

Engine Performance: Heavy Truck, Gasoline Engines

Engine Performance: Import Cars, Light Trucks and Vans

Heating and Air Conditioning: Domestic Cars, Light Trucks and Vans

Heating and Air Conditioning: Import Cars, Light Trucks and Vans

Transmission: Domestic Cars and Light Trucks

Transmission: Import Cars, Light Trucks and Vans

Parts and Labor Estimates

These books are used to estimate repair costs and find part(s) assemblies needed. Provides estimated repair times for nearly every automotive repair procedure imaginable. Usually each publisher produces a separate guide for domestic, imports, and trucks and vans.

Chilton Labor Guide Manual

Mitchell's Parts and Labor Estimating Guide

Motor Parts and Labor Guide

Web Sites

AIC Autosite, Beyond the Basics—An Automotive Encyclopedia
www.autosite.com/garage/encyclop/tocdoc.asp
Divided into twenty-one major sections. Provides information on general automotive topics.

CarCare.org
www.carcarecouncil.org
Provides descriptions of auto parts, parts finder, and useful, car care articles.

Subject Headings

- automobiles—air conditioning—maintenance and repair
- automobiles—electric wiring
- automobiles—repair and maintenance
- automobiles—transmission devices, automatic—maintenance and repair
- automobiles, foreign—motors—maintenance and repair
- trucks, foreign—motors—maintenance and repair

Biographies: A Research Guide

A biography is a story about a person written by someone else. By reading about individuals we can learn many things about them and the people and events surrounding their lives. Biographies also allow us to experience historical events from that person's perspective. The resources below will assist beginning researchers and provide a jumping off point for further research. The terms and phrases listed in the subject headings below can be used to search for more materials in the library's catalog and research databases. If you need further assistance, please ask a librarian.

Indexes and Guides to Biographical Sources

Almanac of Famous People. Detroit: Gale Research, 1989– . Biennial. ISSN: 1040127X.
Includes both historical and contemporary figures with chronological, geographic index and occupational index.

ARBA Guide to Biographical Resources, 1986–1997 edited by Robert L. Wick and Terry Ann Mood. Englewood, CO: Libraries Unlimited, 1998. IBSN: 1563084538.
Guide to selected biographical dictionaries and directories. Each entry gives complete bibliographic information, price, and critical evaluation.

Biography Index. New York: H. W. Wilson, 1946– . Annual. Quarterly updates. ISSN: 00063053.
A cumulative index to biographical material about people from all countries, periods, and occupations (limited to English language books and periodicals).

International Biographical Sources

Current Biography Yearbook. New York: H. W. Wilson, 1940– . Annual. ISSN: 00849499.
Source for biographical data on contemporary people from all professions. A cumulated index exists for the years 1940 through 1995; the 2000 edition indexes the years 1991 through 2000.

International Who's Who. London: Europa, 1935– . Annual. ISSN: 00749613.
Annual that provides succinct biographical data for notables from around the world.

International Who's Who of Women. London: Europa, 1992– . Annual. ISSN: 09653775.
Brief entries on notable women.

Who's Who in the World. Chicago: Marquis, 1972– . Biennial. ISSN: 00839825.
Biennial provides brief biographical data for the world's notables.

Women in World History: A Biographical Encyclopedia edited by Anne Commire and Deborah Klezmer. 16 vols. Waterford, CT: Yorkin, 1999–2000. ISBN: 078763736X.
Biographical essays on women of achievement representing all times and places.

American Biographical Sources: Current

Who's Who Among Black Americans. Detroit: Gale Research, 1976– . Biennial. ISSN: 03625753.
This publication is issued at irregular intervals; the biographies themselves supplied the information.

Who's Who in America. Chicago: Marquis, 1900– . Annual. ISSN: 00839396.
Biennial is the leading source for information on contemporary Americans. Biographical data is supplied by the biographees themselves.

AMERICAN BIOGRAPHICAL SOURCES: RETROSPECTIVE

American National Biography edited by John A. Garraty and Mark C. Carnes. 24 vols. New York: Oxford University Press, 1999. ISBN: 0195206355.
Contains scholarly articles on noted Americans with bibliographies. Indexes by subject, contributor, place of birth, occupation and realm of renown.

National Cyclopaedia of American Biography. 63 vols. New York: James T. White, 1892– . Irregular.
Since this dictionary is not alphabetically arranged, access to the entries is through the general index. While the focus of this work is on the deceased, volumes entitled "current" or "lettered" include those living at the time of compilation. Portraits are supplied throughout the set.

Notable Americans: What They Did, from 1620 to the Present edited by Linda S. Hubbard. Detroit: Gale Group, 1988. ISBN: 0810325349.
Chronological and organizational listings of leaders in government, the military, business, labor, religion, education, cultural organizations, philanthropy, and national associations, including recipients of significant awards and honors.

Who Was Who in America. Chicago: Marquis, 1942– . Annual. ISSN: 01468081.
An index, arranged by name, covers the *Historical Volume* as well as its supplements.

WEB SITES

Biographical Dictionary
www.s9.com/biography/
Dictionary of notable men and women, Searchable by name, birth years, death years, positions held, professions, literary and artistic works, and achievements.

Biography.com
www.biography.com
Companion Web site to A&E's "Biography." Features short biographies of about 25,000 people.

Lives, the Biography Resource
http://amillionlives.com
Includes links to thousands of biography sites on the Web.

SUBJECT HEADINGS

- biography—dictionaries
- biography—indexes
- biography—periodicals

Biology: A Research Guide

Biology is the study of living things and their vital processes, including all of the physicochemical aspects of life. Biology encompasses a broad range of fields of study. For example, marine biologists investigate life in the ocean while ornithologists study birds. Traditionally, the field is split into two area, zoology (study of animals) and botany (study of plants). Researchers will find many authoritative materials listed below that can assist them in beginning their research and fact-finding and locate more in-depth resources. The terms and phrases listed in the subject headings below can be used to search for more materials in the library's catalog and research databases. If you need further assistance, please ask a librarian.

Reference Books

Biology Data Book, 2nd ed. edited by Philip L. Altman and Dorothy D. Katz. 3 vols. Bethesda, MD: Federation of American Societies for Experimental Biology, 1972–1974.

Concise Oxford Dictionary of Zoology edited by Michael Allaby. New York: Oxford University Press, 1992. ISBN: 0192860933.
Entries are under Latin names, with common names in parentheses, and cross-referenced.

The Encyclopedia of the Biological Sciences, 2nd ed. edited by Peter Gray. New York: Van Nostrand Reinhold, 1980. ISBN: 0898743265.
Includes 800 articles on topics in biological sciences.

Henderson's Dictionary of Biological Terms, 12th ed. edited by Eleanor Lawrence. New York: Prentice Hall, 2000. ISBN: 0582414989.
Defines 16,500 terms from biology, botany and zoology with pronunciation.

Synopsis and Classification of Living Organisms edited by Sybil P. Parker. 2 vols. New York: McGraw-Hill, 1982. ISBN: 0070790310.
A comprehensive source that classifies and describes all living organisms. The index includes 35,000 scientific and common names.

Using the Biological Literature: A Practical Guide, 3rd ed. by Diane Schmidt, Elisabeth B. Davis, and Pamela F. Jacobs. New York: Marcel Dekker, 2002. ISBN: 0824706676.
A comprehensive list of sources with an emphasis on current materials.

Washington, Federation of American Societies for Experimental Biology [1964]
This classic handbook provides basic data used in biological research.

Web Sites

The Big Picture Book of Viruses
www.virology.net/Big_Virology/
A catalog of virus pictures. Listed according to the family to which they have been assigned by the International Committee on Taxonomy of Viruses.

BioTech: Life Sciences Resources and Reference Tools
http://biotech.icmb.utexas.edu
This Web site serves everyone from high school students to professional researchers. Includes online dictionary, list of Web sites, career information, and more.

FirstGov for Science
www.science.gov
Gateway to information resources at the U.S. government science agencies. Includes section on biology.

INFOMINE, Scholarly Internet Resource Collection—Biology and Agriculture
http://infomine.ucr.edu/cgi-bin/search?bioag
Gateway developed by the University of California that covers over 2,000 scholarly Internet resources in biology and agriculture.

Scirus
www.scirus.com
Provides searchable database for locating science-specific results on the Web including scientific, scholarly, technical, and medical data.

Associations

American Society for Biochemistry and Molecular Biology
www.asbmb.org/ASBMB/site.nsf
Nonprofit scientific and educational organization with over 11,900 members who teach and conduct research at colleges and universities.

The American Society for Cell Biology
www.ascb.org
Founded in 1960 in order to bring the varied facets of cell biology together and to provide for the exchange of scientific knowledge in the area of cell biology.

American Society for Microbiology
www.asm.org
Advances microbiological sciences through the pursuit of scientific knowledge and dissemination of the results of fundamental and applied research.

Society for Developmental Biology
http://sdb.bio.purdue.edu
This organization's goal is to further the study of development in all organisms and at all levels.

The Society for In Vitro Biology
www.sivb.org
Founded in 1946, as the Tissue Culture Association, in order to foster exchange of knowledge of in vitro biology of cells, tissues, and organs from both plant and animals.

Subject Headings

- adaptation (biology)
- aquatic biology
- biological diversity
- biology
- biology—handbooks, manuals, etc.
- biology, experimental
- botany
- ecology
- genetics
- human biology
- life sciences
- microbiology
- molecular biology
- population biology

BIOMES: A RESEARCH GUIDE

A biome is a plant and animal community that covers a large geographical area that is determined by climate. Plants and animals have features and characteristics that allow them to thrive in their home biome. Important land biomes include the tundra, coniferous and deciduous forests, grasslands, savannas, deserts, chaparral, tropical rain and seasonal forests. The material listed below will help you to gather information on biomes and why they are so important to all living creatures. The terms and phrases listed in the subject headings below can be used to search for more materials in the library's catalog and research databases. If you need further assistance, please ask a librarian.

GENERAL RESOURCES

Encyclopedia of Life Sciences, 2nd ed. edited by Anne O'Daly. 13 vols. Tarrytown, NY: Marshall Cavendish, 2004. ISBN: 0761474420.
Illustrated encyclopedia with articles on agriculture, anatomy, biochemistry, biology, genetics, medicine, and molecular biology.

U·X·L Encyclopedia of Biomes edited by Marlene Weigel. 3 vols. Detroit: UXL, 2000. ISBN: 0787637327.
Covers subject of biomes for all types of environments, their characteristics, creation, and continuity.

TYPES OF BIOMES

Coral Reefs by Lesley A. DuTemple. San Diego: Lucent Books, 2000. ISBN: 1560065974.
Provides an overview of coral reefs as an ecosystem. Discusses their creation, functions, and the organisms that inhabit them.

Deserts by Michael Allaby. New York: Facts On File, 2001. ISBN: 0816039291.
Deals with every aspect of deserts, their inhabitants, their food chains and their biome characteristics.

Grassland by April Pulley Sayre. New York: Twenty-First Century Books, 1997. ISBN: 0805028277.
Includes information on biome characteristics, food webs, and animal and plant life.

Lake and Pond by April Pulley Sayre. New York: Twenty-First Century Books, 1997. ISBN: 0805040897.
Discusses the regions, life forms, and functions of a freshwater pond. There are also pages that discuss biomes and food chains specifically.

Mountains and Valleys by Steve Parker and Jane Parker. London: Chrysalis Children's, 2003. ISBN: 1841389056.
Informative resource on mountains and valleys and the animals and plants that live within this ecosystem.

Oceans by Trevor Day. New York: Facts On File, 1999. ISBN: 0816036470.
Examines every aspect of oceans: their inhabitants, their food chains, and their biome characteristics.

The Prairie by Alison Ormsby. New York: Benchmark Books, 1999. ISBN: 0761408975.
Investigates prairie wildlife and ecosystem, includes several pages of food chains.

River and Stream by April Pulley Sayre. New York: Twenty-First Century Books, 1997. ISBN: 0805040889.
Discusses the life forms, functions, biome characteristics, and food webs for rivers and streams.

Seashore by April Pulley Sayre. New York: Twenty-First Century Books, 1997. ISBN: 0805040854.
Investigates the seashore biomes and the life forms to be found there.

Temperate Forests by Michael Allaby. Danbury, CT: Grolier Educational, 1999. ISBN: 0717293475.
Deals with forests, their inhabitants, their food chains, and their biome characteristics.

Tropical Rainforest by Arnold Newman. New York: Checkmark Books, 2002. ISBN: 0816039739.
A look at tropical rainforests from all aspects, including food webs, habitats, and endangerment of the rainforest.

Wetlands by Peter D. Moore. New York: Facts On File, 2000. ISBN: 0816039305.
Investigates all aspects of the wetlands, including the animals that inhabit it and its functions. Contains many helpful charts, tables and pictures.

Web Sites

Food Chains
www.eagle.ca/~matink/themes/Biomes/foodweb.html
Resource for food chains, pictures, and cycles.

The World's Biomes
www.ucmp.berkeley.edu/glossary/gloss5/biome/
An introduction to the major biomes on Earth.

Subject Headings

- biology
- biotic communities
- ecology
- endangered ecosystems
- food chains
- habitat (ecology)
- life sciences
- life zones
- nature—effect of human beings on

Book Reviews: A Research Guide

Book reviews provide summaries, criticism, and evaluations about published works. Reviews normally appear within a year of the date of publication. In addition to the general resources listed below, reviews can be found in specialized indexes such as *Business Periodicals Index*, *Arts and Humanities Citation Index*, and the *Humanities Index*. The terms and phrases listed in the subject headings below can be used to search for more materials in the library's catalog and research databases. If you need further assistance, please ask a librarian.

Review Indexes

Book Review Digest. New York: H. W. Wilson, 1905– . Monthly. ISSN: 00067326.
Indexes reviews appearing in general interest periodicals of the Anglo-American world. Arranged alphabetically by author of the book, with indexes by title and subject.

Book Review Index. Detroit: Gale Research, 1965– . Three times a year. ISSN: 05240581.
Provides access to reviews of books, periodicals, books on tape, and electronic media representing a wide range of popular, academic and professional interests. More than 600 publications are indexed, including journals and national general interest publications and newspapers.

Children's Book Review Index. Detroit: Gale Research, 1975– . Three times a year. ISSN: 01475681.
Contains reviews of books, periodicals, books on tape, and electronic media intended and recommended for children through age 10.

Children's Literature Review. Detroit: Gale Research, 1976– . Semiannual. ISSN: 03624145.
Excerpts from reviews, criticism, and commentary on books for children and young people. There are cumulative indexes to authors and titles in each volume.

Combined Retrospective Index to Book Reviews in Humanities Journals, 1802–1974. 10 vols. Woodbridge, CT: Research, 1982–1984. ISBN: 089235061X; *Combined Retrospective Index to Book Reviews in Scholarly Journals, 1886–1974.* 15 vols. Arlington, VA: Carrollton Press, 1979–1982. ISBN: 084080167X.
Both indexes are arranged alphabetically by author of the book and include title indexes.

Críticas: An English Speaker's Guide to the Latest Spanish Language Titles. New York: Reed Business Information, 2001– . Bimonthly. ISSN: 15356132.
Published by *Publishers Weekly*, *Library Journal* and *School Library Journal.* Reviews for bookstores, libraries, and publishers.

Index to Book Reviews in the Humanities. Williamston, MI: Thomson, 1960–1990. Annual. ISSN: 00735892.
Indexes scholarly journals from North America and Britain. Arranged alphabetically by author.

Web Sites

Amazon.com
www.amazon.com
Online bookstore that many times includes reviews from *Publishers Weekly*, *Library Journal* and *Library School Journal.*

Bookreporter
www.bookreporter.com
Includes book reviews, author profiles and interviews, excerpts, literary games and contests.

Guardian Unlimited Books: Top 10s
http://books.guardian.co.uk
Reviews of contemporary titles.

H-Net Reviews
www2.h-net.msu.edu/reviews
Reviews from the humanities and social sciences.

London Review of Books
www.lrb.co.uk
London Review of Books features an extensive, full-text archive.

The New York Review of Books
http://nybooks.com/nyrev
Contains select full-text articles.

The New York Times on the Web: Books
www.nytimes.com/books
Includes current reviews, booklists, and a searchable database from 1980.

Salon.com: Books
www.salon.com/books/index.html
Includes daily book reviews, interviews, and a weekly list of author recommendations.

Village Voice Literary Supplement
www.villagevoice.com/vls/
Back issues from October 1998.

Subject Headings

- (author name)—criticism
- book review and title
- books—reviews—indexes
- children's literature—history and criticism—periodicals
- genre—criticism
- humanities—book reviews—indexes

Business Plans: A Research Guide

A business plan defines your business and identifies goals. Many times it is required when applying for business loans. A business plan should answer the following questions: what service or product does your business provide and what needs does it fill, who are the potential customers and how will you reach them, and where will you get the financial resources to start your business? The materials listed below will help you get started. The terms and phrases listed in the subject headings below can be used to search for more materials in the library's catalog and research databases. If you need further assistance, please ask a librarian.

Guides and Handbooks

Anatomy of a Business Plan: A Step-by-step Guide to Building a Business and Securing your Company's Future, 5th ed. by Linda Pinson. Chicago: Dearborn Trade, 2001. ISBN: 0793146003.
Provides a step-by-step process for developing a polished, professional, and results-oriented plan.

Business Plans Handbook, 10th ed. Detroit: Gale Research, 2004. ISBN: 0787664855
Includes sample business plans and templates from various industries. Business plans include type of business; statement of purpose; executive summary; business/industry description; market; product and production; management/personnel; and financial specifics.

How to Write a Business Plan, 6th ed. by Mike P. McKeever. Berkeley, CA: Nolo Press, 2002. ISBN: 0873378636.
Instructions on making realistic financial projections, developing effective marketing strategies, and refining your overall business goals.

Your First Business Plan: A Simple Question and Answer Format Designed to Help You Write Your Own Plan, 4th ed. by Joseph A. Covello and Brian J. Hazelgren. Naperville, IL: Sourcebooks, 2002. ISBN: 1402200021.
Instructional guidelines for those writing their first business plan.

Market Analysis

Market Share Reporter: An Annual Compilation of Reported Market Share Data on Companies, Products, and Services. Detroit: Gale Research, 1991– . Annual. ISSN: 10529578.
Provides an overview of companies, products, and services. Each entry features a descriptive title; data and market description; a list of producers/products along with their market share.

World Market Share Reporter: A Compilation of Reported World Market Share Data and Rankings on Companies Products and Services, 5th ed. Detroit: Gale Research, 2001– . Biennial. ISBN: 0787656585.
Includes 1,670 entries which cover 360 geographic locations. Provides world market share data and rankings on companies, products, and services.

Industry Analysis

Encyclopedia of American Industries, 4th ed. Detroit: Gale Research, 2004. ISBN: 0787690635.
Includes industry description and statistical data for service and non-manufacturing industries (vol. 1) and manufacturing industries (vol. 2).

Standard and Poor's Industry Surveys. New York: Standard & Poor's, 2002– . Quarterly. ISSN: 01964666.
Includes industry profiles and trends, key industry ratios and statistics. A comparative company analysis lists revenues, net income, profit ratios, balance sheet ratios, equity ratios, and per-share data.

U.S. Industry & Trade Outlook. New York: DRI/McGraw-Hill; Washington, DC: U.S. Department of Commerce and the International Trade Administration, 2000– . Annual.

Includes historical data on shipments, imports, exports, employment, industry trends, technology, international competition, forecasts, trade patterns, and major country markets.

Financial Planning

Annual Statement Studies: Financial Ratio Benchmarks and Annual Statement Studies: Industry Default Probabilities and Cash Flow Measures. Philadelphia: Robert Morris Associates, 1923– . Annual.

Provides financial ratios on over 600 industries. For a definition of ratios see *www.rmahq.org/Ann_Studies/ss_faq.html.*

Industry Norms and Key Business Ratios. Murray Hill, NJ: Dun & Bradstreet. Annual. ISSN: 87552396.

Contains industry balance sheet and income statement information with ratios organized by SIC codes.

Web Sites

1997 Economic Census
www.census.gov/epcd/www/econ97.html

Provides a detailed portrait of the U.S. economy once every 5 years. Data can be used to gauge the competition, calculate market share, locate business markets, assist in site location, design sales territories, and set sales quotas.

United States Small Business Administration
www.sba.gov/starting_business/planning/basic.html

Provides a sample outline of a business plan including links to sample plans.

Subject Headings

- business planning
- industries—United States—forecasting
- market share—statistics
- marketing—statistics
- new business enterprises—finance
- new business enterprises—planning
- proposal writing in business
- small business—planning

Buying a Car: A Research Guide

Need to buy or sell a car? Are you dreading fending off hungry salesmen at your local dealership? The following Web sites offer a variety of information to help put both buyers and sellers at ease. Get free no-hassle instantaneous quotes or compare two models side by side. View videos displaying car interiors or read test drive reports. There are plenty of automotive Web sites, but we have narrowed it down to just a few of the best. Buckle up and safe driving! The terms and phrases listed in the subject headings can be used to search for more materials in the library's catalog and research databases. If you need further assistance, please ask a librarian.

Appraisals: What's It Worth?

Edmunds.com
www.edmunds.com
Lists values for new, used, and certified pre-owned vehicles.

Kelley Blue Book
www.kbb.com
Find values of new and used vehicles, but also for snowmobiles, watercraft, and motorcycles. Online version does not include original list price for used vehicles.

NADA Guides
www.nadaguides.com
Includes values for automobiles, boats, manufactured housing, travel trailers, fifth wheel trailers, limousines, and airplanes.

Evaluations

ConsumerGuide
www.consumerguide.com
Offers information on new and used vehicles and includes road tests, photo gallery, competitors, price calculator, specs, and dealer locator.

Intellichoice Car Center
www.intellichoice.com
Provides side-by-side comparisons, safety and recall reports, rebates, and incentives. Highlights best values for new and used cars. Lists lease award winners. Also publishes print title, *The Complete Car Cost Guide.*

Purchase

Autobytel.com
www.autobytel.com
Pre-owned and new autos for sale including auctions. Finance and insure your new vehicle or view 360 degree panoramic interior views.

AutoTrader.com
www.autotrader.com
Search for a used automobile by make, model, and zip code or place an advertisement to sell your own car.

Autoweb.com
www.autoweb.com
Allows side-by-side comparison of vehicle and ability to save your searches. Purchase, sell, finance, insure, and maintain your vehicle at this Web site.

Carpoint.com
http://autos.msn.com
Search or browse by category new and used vehicles: passenger, sports, luxury cars and vans, or trucks. Search geographic radius for used vehicles. Allows side-by-side comparisons. Features new and used car pricing, reviews and buying information.

Cars.com
www.cars.com
Neat feature alert—pick the color of your car and watch it change instantaneously. E-mail reminder service, automobile advice, and comparison shopping, side-by-side.

CarsDirect
www.carsdirect.com
Offers brochures on new cars including vehicle specs, reviews, 360 degree interior shots from the driver's seat, photo gallery, handling and safety information.

Subject Headings

- automobiles—cost of operation—periodicals
- automobiles—prices—periodicals
- automobiles—purchasing—periodicals
- trucks—cost of operation—periodicals
- trucks—prices—periodicals
- trucks—purchasing—periodicals
- vans—cost of operation—periodicals
- vans—prices—periodicals
- vans—purchasing—periodicals

Career Information and Job Hunting: A Research Guide

Job seekers looking for new career opportunities will find the resources listed below a great starting point for their job hunt. Resources include career exploration for those just entering the workforce to midlife career changes for experienced workers. Researchers will also find answers to some of the more difficult questions such as matching your college major and preferred skills to an occupation. Once a field or job title is identified career exploration leads to identifying potential employers or sources for job advertisements in the chosen field. Librarians are also available to help guide the way to uncover even more sources of information including the education that may lead to a degreed or certified position or specialized directories of company information. The terms and phrases listed in the subject headings below can be used to search for more materials in the library's catalog and research databases. If you need further assistance, please ask a librarian.

Books

Guide to Internet Job Searching 2004–2005 by Margaret Riley Dikel. New York: McGraw-Hill, 2004. ISBN: 007141374X.
Offers expert advice on how to find and use Internet resources to run a successful online job hunt. Includes use of bulletin boards, job listings, recruiter information, discussion groups, and resume posting services.

What Color Is Your Parachute?, 2004: A Practical Manual for Job-Hunters and Career-Changers by Richard Nelson Bolles. Berkeley, CA: Ten Speed Press, 2003. Annual. ISBN: 1580086152.
Completely revised edition. Provides a step-by-step plan for finding meaningful work.

Career Exploration: Many Occupations

America's Career InfoNet
www.acinet.org
Find wages and employment trends, occupational requirements, state by state labor market conditions, millions of employer contacts nationwide, and the most extensive career resource library online.

Career Guide to Industries
http://stats.bls.gov/oco/cg/
Provides information on available careers by industry, including the nature of the industry, working conditions, employment, occupations in the industry, training and advancement, earnings and benefits, employment outlook, and lists of organizations that can provide additional information.

Major Resource Kits, MBNA Career Services Center, University of Delaware
www.udel.edu/CSC/mrk.html
These kits help answer the question, "What do I do with a major in…" Matches college majors with relevant job titles.

Occupational Outlook Handbook. Washington, D.C.: U.S. Department of Labor, Bureau of Labor Statistics, biennial.
www.bls.gov/oco/
Handbook describes what workers do on the job as well as working conditions, the training and education needed, earnings, and expected job prospects in a wide range of occupations.

O*Net OnLine
http://online.onetcenter.org
Find occupations by title, classification, and by skill.

Career Exploration: Specific Occupations

Associations on the Net, The Internet Public Library
www.ipl.org/div/aon/
Search for associations in your area of interest. Trade associations in a particular subject area usually offer career information.

Careers in Real Estate, Realtor.org
www.realtor.org/realtororg.nsf/pages/careers
Provides brief information on various careers in real estate including information on where to find more.

MedCareers
www.medcareers.com
Web site about medical, nursing, and healthcare industry.

Nursing Careers, NurseWeek.com
www.nurseweek.com/career/
Provides brief overview of specific jobs in the nursing field.

Today's Military
www.todaysmilitary.com
Provides background information on all types of military careers.

Locating a Potential Employer

Many job hunters interested in particular jobs or jobs in certain industries can look to many resources in order to identify employers that hire workers in their area of interest. Once identified, job seekers can continue to monitor the employer's Web site or job line for upcoming employment opportunities. Specialized directories, business directories, and even the local yellow pages can help identify potential employers. State Employment Officers often have lists of the state's largest employers.

Web Sites

America's Job Bank
www.ajb.dni.us
Lists jobs from 1,800 State Employment Service offices nationwide.

CareerBuilder
www.careerbuilder.com
Post resumes, search job listings, and more.

Monster Board
www.monster.com
Offers a wide range of career-related resources including job, career, and networking information.

The Riley Guide
www.rileyguide.com
A very thorough and current guide to job hunting and career information. Includes a comprehensive list of job banks and recruiting sites.

Subject Headings

- career changes
- career development
- career plateaus
- educational counseling
- employment
- employment in foreign countries
- job hunting
- job hunting—computer network resources
- occupations
- vocational guidance

Celebrations Around the World: A Research Guide

Learn how different cultures and people celebrate holidays and festivals throughout the world. The materials listed below will get you started in your research to understand more about religious, secular, legal, public, and national celebrations and traditions around the world. The terms and phrases listed in the subject headings below can be used to search for more materials in the library's catalog and research databases. If you need further assistance, please ask a librarian.

Reference Books

American Book of Days, 4th ed. by Stephen G. Christianson. New York: H. W. Wilson, 2000. ISBN: 0824209540.
Essays explore significant events for each day of the year including military, scientific, ethnic, and cultural events. An appendix features historical documents.

Chase's Calendar of Events. New York: McGraw-Hill, 2004– . Annual. ISBN: 0071424059.
A calendar containing over 12,000 listings including special days, weeks, and months as well as holidays, historical anniversaries, fairs, and festivals.

Encyclopedia of Christmas & New Year's Celebrations. Detroit: Omnigraphics, 2000. ISBN: 0780803876.
Over 240 entries covering Christmas, New Year's, and related days of observance, including folk and religious customs, history, legends, and symbols worldwide.

Festivals of the World: The Illustrated Guide to Celebrations, Customs, Events, and Holidays edited by Elizabeth Breuilly, et al. New York: Checkmark Books, 2002. ISBN: 0816044813.
Includes Jewish, Christian, Muslim, Hindu, Buddhist, Sikh, Taoist, Zoroastrian, Shinto, Jain, Baha'i, and Rastafarian festivals.

The Folklore of American Holidays. Gale Research, 1987. ISBN: 0810388642.
Descriptions of origin, historical background, and general characteristics of more than 125 holidays celebrated in America. Chronologically arranged, from New Year's Day through Christmas, entries include source and bibliographical information.

The Folklore of World Holidays, 2nd ed. Gale Research, 1998. ISBN: 0810389010.
Lists the date of the holiday, description of the holiday, and traditions and folklore associated with it for every country that celebrates that particular holiday.

Holidays, Festivals, and Celebrations of the World Dictionary, 3rd ed. edited by Helene Henderson and Sue Ellen Thompson. Detroit: Omnigraphics, 2004. ISBN: 0780804228.
A reference guide to popular, ethnic, religious, national, and ancient holidays including nearly 2,500 observances from all 50 states and more than 100 nations.

Junior Worldmark Encyclopedia of World Holidays by Robert H. Griffin and Ann H. Shurgin. 4 vols. Detroit: UXL, 2000. ISBN: 0787639273.
Vol. 1: Buddha's birthday, Carnival, Christmas; vol. 2: Easter, Halloween and festivals of the dead; vol. 3: Hanukkah, Independence Day, Kwanzaa; vol. 4: New Year, Ramadan and 'Id al-Fitr, Thanksgiving and harvest festivals.

World Holiday, Festival & Calendar Books edited by Tanya Gulevich. Detroit: Omnigraphics, 1998. ISBN: 0780800737.
Features more than 1,000 holiday-related resources, including dictionaries, guidebooks and reference sources, folklore and ethnic studies, children's books, and historical background studies.

Web Sites

Earth Calendar
www.earthcalendar.net
Provides easy access to holidays by date, country, and religion. Includes brief information with links to related Web sites.

HistoryChannel.com: The History of the Holidays
www.historychannel.com/exhibits/holidays/main.html
Provides history and traditions of Christmas, Kwanzaa, and Hanukkah.

Infoplease: Calendar and Holidays
www.infoplease.com/ipa/A0001832.html
Lists calendars for major holidays, both religious and secular.

Turkey for the Holidays, University of Illinois
www.urbanext.uiuc.edu/turkey/
A guide to selection, cooking, and carving Thanksgiving dinner. Includes additional Thanksgiving links.

Subject Headings

- fasts and feasts—country
- festivals—country
- festivals—dictionaries
- holidays—country
- holidays—dictionaries

Chemistry and Chemical Engineering: A Research Guide

The field of chemistry deals with the properties, composition and structure of substances, their reactions and transformations, and the energy released or absorbed during those processes. Chemical engineers study chemistry, physics, and mathematics to further understand chemical reactions and their effects. The materials below will help beginning researchers as well as those looking for more comprehensive coverage. The terms and phrases listed in the subject headings below can be used to search for more materials in the library's catalog and research databases. If you need further assistance, please ask a librarian.

Reference Books

Chemical Abstracts. Columbus, OH: American Chemical Society, 1907– . Weekly. ISSN: 00092258.
Weekly publication of indexes and abstracts for international chemical literature including journals and conference proceedings.

Chemical Technicians' Ready Reference Handbook, 4th ed. edited by Gershon J. Shugar and Jack T. Ballinger. New York: McGraw-Hill Professional, 1996. ISBN: 0070571864.
Provides information on techniques and procedures, measuring temperatures, heating and cooling.

Concise Chemical and Technical Dictionary, 4th ed. by Harry Bennett. New York: Chemical, 1986. ISBN: 0820602043.
Includes over 50,000 definitions. Brevity is the rule, with definitions being no more than four lines.

CRC Handbook of Chemistry and Physics. Cleveland: CRC Press, 1913– . Annual. ISSN: 01476262.
Contains basic chemical and physical data for compounds, as well as many other useful tables, constants and formulas, and definitions.

Encyclopedia of Chemical Technology, 5th ed. New York: John Wiley & Sons, 2004. ISBN: 0471485187.
Covers preparation, manufacturing processes, uses of materials, properties of chemical substances, and chemistry fundamentals.

The Facts On File Dictionary of Chemistry, 3rd ed. edited by John Daintith. New York: Facts On File, 1999. ISBN: 0816039097.
Handy dictionary of terms and concepts.

Lange's Handbook of Chemistry, 15th ed. edited by John A. Dean. New York: McGraw-Hill, 1999. ISBN: 1591241138.
Contains sections on mathematics, general information and conversion tables, atomic and molecular structure, inorganic chemistry, organic chemistry, analytical chemistry, electrochemistry, spectroscopy, and thermodynamic and physical properties of compounds.

The Merck Index: An Encyclopedia of Chemicals, Drugs, and Biologicals, 13th ed. edited by Maryadele J. O'Neil, et al. Whitehouse Station, NJ: Merck, 2001. ISBN: 0911910131.
Description of over 10,000 chemicals, drugs, and biological substances. Entries include name, synonyms and trademarks, molecular formulas, molecular weight, structure, physical data, and annotated literature references. Indexed by name, registry number, therapeutic category and biological activity, and formula.

WEB SITES

ChemFinder.com
http://chemfinder.cambridgesoft.com
Search by chemical name, CAS Number, molecular formula, or molecular weight to obtain information about a particular chemical (physical properties, health information, and uses).

Chemical Periodic Table
www.chemicool.com
Information covered includes general states, energies, oxidation and electrons, appearance and characteristics, reactions, other forms, radius, conductivity, and abundance.

ChemIndustry.com
www.neis.com
A search engine for chemical and related industry professionals, engaged in every discipline, from discovery through development, manufacturing and marketing.

Rolf Claessen's Chemistry Index
www.claessen.net/chemistry/
Collection of links covering all areas of chemistry.

WebElements Periodic Table
www.webelements.com
Collection of property data on elements.

ASSOCIATIONS

The American Chemical Society
www.chemistry.org
Covers all fields of chemistry. Site includes membership information, events and conferences, publications.

American Institute of Chemical Engineers (AICHE)
www.aiche.org
Professional association of more than 50,000 members that provides leadership in advancing the chemical engineering profession.

SUBJECT HEADINGS

- chemistry—dictionaries
- chemistry—encyclopedias
- chemistry—manipulation—handbooks, manuals, etc.
- chemistry—tables—periodicals
- chemistry, technical—encyclopedias
- physics—tables—periodicals
- technology—dictionaries
- technology—encyclopedias

College Search: A Research Guide

A number of excellent books and Web sites provide information for college-bound students and their parents. Listed here are some starting points for gathering information on testing, the college application process, sources of financial aid, school rankings, and contact information. Prospective students may also want to review some of the career materials to explore various occupations and their educational requirements. The terms and phrases listed in the subject headings below can be used to search for more materials in the library's catalog and research databases. If you need further assistance, please ask a librarian.

Web Sites

2004 Colleges, College Scholarships, and Financial Aid Page
www.college-scholarships.com
A directory of college and university admission office e-mail addresses, arranged by state, plus supplemental links to financial aid offices, and information on tests such as GRE, GMAT, TOEFL, and homepages of more than a thousand colleges and universities.

CollegeBoard
www.collegeboard.com
Preparing for and taking the SATs, registration dates and applications, general tips on getting ready for college.

College Calculators, CollegeBoard.com
www.collegeboard.com/article/0,1120,6-29-0-401,00.html
Use these calculators to estimate college costs, family contribution, and monthly repayment obligation for varying loan amounts.

Education, *U.S. News & World Report*
www.usnews.com/usnews/edu/eduhome.htm
Sponsored by U.S. News Online, this Web site offers college and graduate school rankings and additional features such as searching for scholarships by category, listings of internships, and "hot job tracks."

FAFSA and Title IV School Code List, Federal Student Aid
http://studentaid.ed.gov
Obtain either the electronic or the print version of the "Free Application for Federal Student Aid," and the "Title IV School Code List" from the U.S. Department of Education's Web site.

IPEDS College Opportunities online, National Center for Education Statistics
http://nces.ed.gov/IPEDS/cool/index.asp
A U.S. Department of Education site designed to help you obtain information about colleges. Schools are searchable by location, program of study, size, or a combination of factors.

Directories

The Fiske Guide to Colleges by Edward B. Fiske, et al. New York: Times Books, 1989– . Annual. ISSN: 10427368. Previous title: *Selective Guide to Colleges*, 1983–1988.
In addition to the usual statistical information, this annual guide provides median test scores, number of applicants and percent accepted, as well as ratings on quality of academics and quality of life.

Gourman Report by Jack Gourman. Los Angeles: National Education Standards, 1980– . Biennial. ISSN: 10497188.
Ratings of undergraduate programs in American and international universities. Assigns a precise, numerical score to each school and undergraduate program listed.

Index of Majors and Graduate Degrees. New York: College Entrance Examination Board, 1980– . Annual. ISSN: 15416828.
Helps students determine which of the six hundred majors is right for them and which colleges offer that major.

Peterson's Guide to Four Year Colleges. Lawrenceville, NJ: Peterson's, 1984– . Annual. ISSN: 15442330.
Data profiles for more than 2,100 institutions—listed alphabetically by state, followed by other countries.

Peterson's Guide to Two Year Colleges. Lawrenceville, NJ: Peterson's. Annual. ISSN: 08949328.
Information on accredited two-year undergraduate institutions in the U.S. Includes detailed two-page descriptions written by admissions personnel for nearly 70 colleges.

Rugg's Recommendations on the Colleges by Frederick E. Rugg. East Longmeadow, MA: Celecom, 1980– . Annual. ISBN: 1883062489.
Identifies recommended undergraduate programs in various fields at 660 colleges. Includes profiles of these schools.

Financial Aid

College Board College Cost and Financial Aid Handbook. New York: College Entrance Examination Board. Annual. ISBN: 0874476976
Provides the facts and figures needed to calculate the true costs at over 2,700 four-and two-year colleges. It includes itemized tuition and fee information; itemized charts of all student expenses; payment plan information; indexes of colleges offering merit, academic, and athletic scholarships.

The following biennial directories list scholarships, fellowships, loans, grants, awards, and internships open primarily to a specific audience.

- *Directory of Financial Aid for Women.* ISSN: 07325215.
- *Financial Aid for African Americans.* ISSN: 1099906X.
- *Financial Aid for Asian Americans.* ISSN: 10999124.
- *Financial Aid for Hispanic Americans.* ISSN: 10999078.
- *Financial Aid for Native Americans.* ISSN: 10999116.

Peterson's College Money Handbook, 21st ed. Lawrenceville, NJ: Thomson/Peterson's, 1984– . Annual. ISBN: 0768912342.
Annually updated reference guide to more than 1,600 individual colleges' student financial aid appropriations. Contains a chart that shows actual, average costs of specific colleges.

Scholarships, Fellowships, and Loans by S. Norman Feingold and Marie Feingold. Boston: Bellman, 1949– . Annual. ISBN: 0787634751.
Provides more than 3,700 sources of education-related financial aid and awards at all levels of study.

Subject Headings

- degrees, academic—United States—directories
- fellowships and scholarships—United States—directories
- fellowships and scholarships—United States—periodicals
- scholarships—United States—periodicals
- universities—Canada—directories
- universities—United States—directories
- universities and colleges—United States—curricula—directories

Company Information: A Research Guide

Researchers often search for company information by characteristics such as amount of employees or sales, geographic location, industry category. They may also have a variety of reasons to locate companies such as employment opportunities or leads generation. Researchers will find it more difficult to locate information about private companies simply because they are not required to report various filings with the Securities Exchange Commission; however, there are some sources that list privately held companies, as well as information that can be found on a company's Web site, article searches, local business press, or Secretary of State filings. The terms and phrases listed in the subject headings below can be used to search for more materials in the library's catalog and research databases. If you need further assistance, please ask a librarian.

Print Directories

Hoover's Handbook of American Business 2004, 14th ed. Austin, TX: Hoover's Business Press, 2003. Annual. ISBN: 1573110884.
Two-volume set contains in-depth profiles of 750 of America's largest and most influential companies. Examines the personalities, events, and strategies that have made these enterprises leaders in their fields.

Hoover's Handbook of Emerging Companies 2003, 10th ed. Austin, TX: Hoover's Business Press, 2003. Annual. ISBN: 1573110833.
Covers selected U.S. public companies with sales between $10 million and $1 billion, at least 3 years of reported sales, at least a 30% rate of sales growth during that time, and positive net income for their most recent fiscal year.

Hoover's Handbook of Private Companies, 9th ed. Austin, TX: Hoover's Business Press, 2004. Biennial. ISBN: 1573110906.
Covers 900 nonpublic U.S. enterprises including large industrial and service corporations, hospitals and health care organizations, charitable and membership organizations, mutual and cooperative organizations, joint ventures, government-owned corporations, and major university systems.

Hoover's Handbook of World Business, 10th ed. Austin, TX: Hoover's Business Press, 2004– . Annual. ISBN: 1573110892.
In-depth profiles of 300 of the most influential firms from Canada, Europe, and Japan, as well as companies from the fast-growing economies of such countries as China, India, and Taiwan.

Mergent Industrial Manual. New York: Mergent FIS, 1909– . Annual. ISSN: 15400832.
Covers 2,000 top industrial corporations. Includes history, business, properties, subsidiaries, officers, directors, long-term debt, capital stock, letter to shareholders, notes to financial statements, and financial data.

Standard & Poor's Register of Corporations, Directors and Executives. New York, Standard & Poor's, 1973– . Annual. ISSN: 03613623.
A principal source for company identification indexed by industrial code, geographical location, and by executive and directors' names.

Rankings and Lists

Fast Companies Database
http://fcke.fastcompany.com
World's most interesting and innovative companies.

Fortune's Index of Lists
www.fortune.com/fortune/alllists
Includes well-known lists like Fortune 500, as well as others like the Global 500, America's Most Admired Companies, World's Most Admired Companies, 100 Best Companies to Work For and many more.

Scoreboards, *BusinessWeek*
www.businessweek.com/common/tools.htm
Provides interactive lists that rank companies by certain investment variables.

Telephone Directories: Yellow Pages, White Pages, and Reverse Lookups

Many directories include search by person's or company's name, yellow page heading, or reverse telephone number. These directories may provide postal addresses, telephone numbers and e-mail addresses for people.

- 411locate: *www.411locate.com*
- AnyWho online Directory: *www.anywho.com*
- Infobel Worldwide Directories: *www.infobel.com/teldir/*
- InfoSpace: *www.infospace.com*
- InfoUSA: *www.infousa.com* (see Free Yellow Pages)
- International Directories, WhitePages.com: *www.whitepages.com/intl_sites.pl*
- The Ultimates: *www.theultimates.com*

Subject Headings

- business enterprises—directories
- business enterprises—(name of state)—directories
- commerce—(name of country)—directory
- commerce—directories
- corporations—directories
- corporations—(name of country)—directories
- corporations—(name of country)—finance—directories
- corporations—(name of country)—rankings—directories
- directors of corporations—directories
- financial management—(name of country)—directory
- industries—states—directories
- industries—(name of country)—directories
- industry—(name of country)—directory
- manufactures—(name of state)—directories
- manufacturing industries—(name of state)—directories
- service industries—(name of state)—directories

Computer Science and Software Engineering: A Research Guide

As the field of computers and software has developed and evolved, its body of literature has grown. The resource materials listed below will help beginning researchers become acquainted with some of the standard materials in this subject area, as well as provide a starting point for more advanced research. The terms and phrases listed in the subject headings below can be used to search for more materials in the library's catalog and research databases. If you need further assistance, please ask a librarian.

Reference Books

Computer Science: An Overview, 7th ed. by J. Glenn Brookshear. Boston: Pearson Addison Wesley, 2003. ISBN: 0201781301.
Provides an overview to the field of computing, including networking and the Internet, software engineering, public-key encryption, and artificial intelligence.

Concepts of Programming Languages, 6th ed. by Robert W. Sebesta. Boston: Pearson Addison Wesley, 2004. ISBN: 0321193628.
Explains how to evaluate existing and future programming languages and how to design compilers.

Concise Encyclopedia of Computer Science by edited by Edwin D. Reilly, et al. New York: Palgrave Macmillan, 2004. ISBN: 0333998685.
A concise and annotated version of the classic work *Encyclopedia of Computer Science*, this volume is an authoritative and user-friendly reference providing clear and concise explanations of the latest technology and its applications, including past, present, and predicted future trends in computer science.

Dictionary of Computer Science, Engineering, and Technology by Phillip A. LaPlante. Boca Raton, FL: CRC Press, 2001. ISBN: 0849326915.
Includes over 8,000 terms and covers major topics like artificial intelligence, programming languages, software engineering, operating systems, database management, and privacy issues.

Encyclopedia of Computers and Computer History edited by Raul Rojas. 2 vols. Chicago: Fitzroy Dearborn, 2001. ISBN: 1579582354.
Covers the history of computing from the abacus to eBay, biographies of major figures in the history of computers, company background, lists of computer terminology, and profiles of pioneering computers.

Encyclopedia of Computer Science, 4th ed. edited by Anthony Ralston, Edwin D. Reilly, David Hemmendinger. Hoboken, NJ: John Wiley & Sons, 2003. ISBN: 0470864125.
Providing explanations of the latest technology and its applications, including past, present, and predicted future trends in computer science.

Encyclopedia of Software Engineering, 2nd ed. edited by John J. Marciniak. New York: John Wiley & Sons, 2002. ISBN: 0471377376.
Includes information on all aspects of engineering for practitioners who design, write, or test computer programs.

The Facts On File Dictionary of Computer Science, 4th ed. edited by Valerie Illingworth and John Daintith. New York: Facts On File, 2001. ISBN: 0816042853.
Defines the most relevant and frequently used terms in modern computer science, including the latest changes and trends in hardware, software, and applications.

MCSE: The Core Exams in a Nutshell: A Desktop Quick Reference by Michael G. Moncur. Cambridge, MA: O'Reilly, 2002. ISBN: 1565927214.
Contains the essence of the information that is required to pass exams for Microsoft Certified Systems Engineers.

Software Engineering, 6th ed. by Ian Sommerville. Boston: Addison-Wesley, 2000. ISBN: 020139815X.
Provides a comprehensive discussion of software engineering techniques and shows how they can be applied in practical software projects. Covers the software process and software process technology, system integration, requirements management, risk analysis, pattern-based reuse, distributed system engineering, and legacy systems.

Associations and Societies

Association for Computing Machinery
www.acm.org
Organization that advances the skills of information technology professionals and students worldwide. Web site contains information about membership, computing literature, publications, and conferences.

Free Software Foundation
www.fsf.org
Its mission is to preserve, protect, and promote the freedom to use, study, copy, modify, and redistribute computer software, and to defend the rights of software users.

IEEE Computer Society
www.computer.org
Leading organization of computer professionals.

Web Sites

ElsevierComputerScience
www.elseviercomputerscience.com
Articles and publications about computer science.

"Computer Science: A Guide to Selected Resources on the Internet" by Michael Knee. *C&RL News*, June 2001, vol. 62, no. 6. Also available online at *www.ala.org/ala/acrl/acrlpubs/crlnews/backissues2001/june1/computerscience.htm*.

ZDNet
www.zdnet.com
Resources for information technology professionals, including news, reports, reviews, technical updates.

Subject Headings

- analog computers
- computer engineering
- computer hardware
- computer science
- digital computers
- hybrid computers
- programming languages
- software

Consumer Information: A Research Guide

This guide is designed to help consumers locate information on the value, pricing, and reliability of goods and services with a focus on consumer advocacy. There are specific headings devoted to higher profile issues in this area such as Children's Products, Automobiles, and Health. Though the main focus of this guide is on items and services that can be purchased in the United States, some links may lead to resources outside of the U.S. The terms and phrases listed in the subject headings below can be used to search for more materials in the library's catalog and research databases. If you need further assistance, please ask a librarian.

General Online Sources

Better Business Bureau (United States and Canada)
www.bbb.org
Provides information on business and consumer alerts, instructs consumers about how to file a complaint online, and includes a directory of BBBs.

Consumer Alert
www.consumeralert.org
Consumer advocacy organization that generates several e-newsletters for free including *Consumer Comments*, *CPSC Monitor*, and *On the Plate*.

ConsumerReview.com
www.consumerreview.com
Free Web site that offers product reviews written by consumers. Searchable and browsable.

ConsumerWorld.org
www.consumerworld.org
Searchable Web site that provides access to hundreds of consumer resources on the Internet. It is categorized according to "hot" consumer topics, such as credit card information, mortgage rates, and travel.

Government Sources

Federal Citizen Information Center (FCIC)
www.pueblo.gsa.gov
One-stop shopping center for federal consumer publications. Their Web site provides a wealth of information on consumer issues, including the Consumer Action Handbook (revised annually) and other FCIC publications.

Federal Trade Commission for the Consumer
www.ftc.gov/ftc/consumer.htm
Provides consumer protection information on such topics as lending, automobiles, advertising, children's issues, products and services, energy and the environment, travel, seniors, and many more.

FirstGov for Consumers
www.consumer.gov
Directory to government-generated, consumer information with the following categories: Food, Product Safety, Health, Home and Community, Money, Transportation, Children, Careers and Education, Technology, and Miscellaneous.

Public Education and Information, Attorneygeneral.gov
www.attorneygeneral.gov/pei/bcp.cfm
Displays information on the Puppy Lemon Law, telephone scams, weight loss products, credit advice, advice for seniors, and much more.

U.S. Consumer Product Safety Commission
www.cpsc.gov
An Independent Federal Regulatory Agency with a primary focus on family safety. The agency works with industry to develop product standards and bans products where there are no standards, but high risk. Look here for important information on product recalls.

The U.S. Consumer Product Safety Commission provides two major reports relating to kids and toys:

Toy Hazard Recalls
www.cpsc.gov/cpscpub/prerel/category/toy.html
Toy recalls are listed in reverse chronological order. Sponsored by the U.S. Product Safety Commission.

Toy Safety Publications
www.cpsc.gov/cpscpub/pubs/toy_sfy.html
This Web site provides safety tips on specific toys and general information about selecting appropriate toys for children.

Automobiles

Automobile Pricing and Performance
NADA Official Used Car Guide: *www.nadaguides.com*
Kelley Blue Book: *www.kbb.com*

CARFAX
www.carfax.com
Offers vehicle history report for a fee.

Edmund.com
www.edmunds.com
Offers pricing and evaluation on new, used, and certified cars. Includes section on future models.

National Highway Traffic Safety Administration
www.nhtsa.dot.gov/cars/
Includes categories on Safety Problems and Issues, Testing Results, Regulations and Standards, and Research and Development.

National Lemon Law Center
www.nationallemonlawcenter.com
Provides state by state Lemon Law information for used cars and also has a consumer product lawyer locator.

Magazines

Consumers Digest
Consumer Guide
Consumer Reports
Consumer Reports Newsletter on Health
Consumer Reports Newsletter on Travel
Consumers Research

Subject Headings

- consumer advocacy
- consumer education
- consumer information
- consumer protection

Contemporary Issues: A Research Guide

Many times students are asked to write position papers that refute or support particular viewpoints about issues relevant in today's society. Not only students, but others interested in contemporary issues and the various complex policies surrounding them will find the resources below to be helpful. Issues like abortion, cloning, death penalty, euthanasia, and full-day kindergarten pose challenges to people who want to fully understand the major issues at hand. The materials below will help beginning researchers as well as those looking for more comprehensive coverage. The terms and phrases listed in the subject headings below can be used to search for more materials in the library's catalog and research databases. If you need further assistance, please ask a librarian.

Books and Magazines

CQ Researcher. Washington, DC: Congressional Quarterly, 1991. Weekly. ISSN: 10562036.
Weekly reports on wide range of current events and issues; each report on one topic and gives both sides of an argument.

Opposing Viewpoints Series. San Diego: Greenhaven Press.
Over hundreds of topics are included in this series. Each title explores a specific issue by placing expert opinions in a unique pro/con format.

Sample Titles:
- *Abortion.* 2001. ISBN: 0737707771.
- *American Values.* 2002. ISBN: 0737703431.
- *Death Penalty.* 2001. ISBN: 0737707917.
- *Extremist Groups.* 2001. ISBN: 0737706554.
- *Global Warming.* 2001. ISBN: 073770909X.

Web Sites

Debatabase, International Debate Education Association
www.debatabase.org
Includes arguments for and against hundreds of debating topics. Written by expert debaters, judges, and coaches. Also included are background summaries, links to Web sites of interest, and recommended books, example motions, and user comments. Sponsored by the International Debate Education Association.

Justice Talking
www.justicetalking.org
Tackles current hot topics with reports from the field, polling analysis, and compelling debate between the nation's leading advocates and political opposites. Sponsored by National Public Radio.

Public Agenda
www.publicagenda.org
Explains and clarifies public attitudes about complex policy issues. Provides research studies and issue guides on topics like the right to die, illegal drugs, America's global role, and more.

Organizations

International Debate Education Association (IDEA)
www.idebate.org
IDEA is an independent membership organization of national debate programs and associations and other organizations and individuals that support debate. IDEA provides assistance to national debate associations and organizes an annual international summer camp.

Topics to Explore

The topics listed below include some of the most highly debated topics in contemporary times. Students can search the library catalog, informational databases, and search engines using the terms listed below with keywords like position papers, pro and con with the topic name. (For example: abortion and "position paper").

Abolition of boxing
Abortion
Adoption
Affirmative action
AIDS (disease)
Alcoholism
Animal rights
Arranged marriages
Assisted suicide (euthanasia)
Banning of Confederate flag
Beauty contests
Bioethics
Biological weapons
Birth control
Cameras in courtrooms
Capital punishment
Censorship
Chemical weapons
Child abuse
Child curfews
Church and state
Compulsory voting
Condoms in schools
Conservation
Creationism vs. evolution
Divorce
Drinking (alcoholic beverages)
Driving while intoxicated
Drugs and athletes
Election of judges
Emigration and immigration
Endangered species
Environmentalism
Faith schools
Feminism
Flag burning
Gambling
Gangs
Gay marriages
Gays in the military
Genocide
Global warming
Gun control
Health care reform
Homelessness
Homeschooling
Homosexuality
Human cloning
Identity cards
Internet censorship
Invasion of Iraq
Islamic fundamentalism
Juvenile alcoholism
Juvenile drug abuse
Juvenile offenders
Legalization of marijuana
Mandatory retirement
Media violence
Medical care
Medical ethics
Mental disorders
Mental health
Middle East
Minimum wage
Narcotics control
Narcotics legalization
National security
Nuclear weapons
Nutrition
Oceans
Political corruption
Pollution
Population growth
Pornography
Poverty
Prisons
Prohibition of school prayer
Public assistance
Race relations
Racism
Rape
Religion
Renewable energy
Right of privacy
School violence
Sex education
Sexual behavior
Single-sex schools
Smoking
Stem cell research
Suicide
Sunday entertainment
Teenage pregnancy
Teenage sexual behavior
Terrorism
United States foreign relations
Use of mobile telephones in cars
Violence
War crimes
Water pollution
Welfare reform
Women's rights
Working women
Youth
Zero tolerance
Zoos

Countries: A Research Guide

Whether your research takes you to a city in your own country or one around the world, the resources listed below will help you identify geographic names, statistical data, business and cultural information, and basic facts like major religions, population, country songs, and symbols. The materials below will help beginning researchers as well as those looking for more comprehensive coverage. The terms and phrases listed in the subject headings below can be used to search for more materials in the library's catalog and research databases. If you need further assistance, please ask a librarian.

Gazetteers

The Columbia Gazetteer of the World edited by Saul Bernard Cohen. New York: Columbia University Press, 1998. ISBN: 0231110405.
Entries include political divisions (states, provinces, capitals), the physical world (oceans, mountains, volcanoes), and special places (military bases, dams, national parks).

Merriam-Webster's Geographical Dictionary, 3rd ed. Springfield, MA: Merriam-Webster, 1997. ISBN: 0877795460.
Contains key data about the countries, cities, and natural features of today's world. More than 48,000 entries and 252 maps provide population, size, economic data, historical notes, and more.

Country Information

Background Notes. Washington, DC: U.S. Department of State.
www.state.gov/r/pa/ei/bgn/
Factual publications that contain information on all countries throughout the world with which the United States has relations including facts on the country's land, people, history, government, political conditions, and economy.

Demographic Yearbook/Annuaire Démographique. United Nations, 1948– . ISSN: 00828041.
Country demographic statistics prepared by the Statistical Division of the United Nations.

Chiefs of State and Cabinet Members of Foreign Governments. Washington, DC: National Foreign Assessment Center. Monthly. ISSN: 01622951.
www.cia.gov/cia/publications/chiefs/index.html
Includes governments of the world, some of them not officially recognized by the United States or with which the United States has no diplomatic exchanges.

Country Studies. Washington, DC: Federal Research Division, Library of Congress.
http://lcweb2.loc.gov/frd/cs/cshome.html
Presents a description and analysis of the historical setting and the social, economic, political, and national security systems and institutions of countries throughout the world. Covers 102 countries and regions.

Travel Warnings and Consular Information Sheets. Washington, DC: U.S. Department of State.
http://travel.state.gov/travel_warnings.html
Issues travel warnings, location of the U.S. Embassy or Consulate, unusual immigration practices, health conditions, minor political disturbances, unusual currency and entry regulations, crime and security information, and drug penalties.

The World Factbook. Washington, DC: Central Intelligence Agency, 1981– . Annual. ISSN: 02771527.
www.cia.gov/cia/publications/factbook/index.html
Covers the following categories: geography, people, government, economy, communications, transportation, military, and transnational issues.

Worldmark Encyclopedia of the Nations, 11th ed. edited by Timothy L. Gall. 6 vols. Detroit: Gale Group, 2003. ISBN: 0787673307.
Presents information on countries and dependencies from around the world. Vol. 1: United Nations; vol. 2: Africa; vol. 3: Americas; vol. 4: Asia and Oceania; vol. 5: Europe; vol. 6: World leaders.

Organizations

UNICEF (United Nations Children's Fund)
www.unicef.org
Provides health care, clean water, improved nutrition, and education to children in Africa, Asia, Central and Eastern Europe, Latin America, and the Middle East.

The United Nations
www.un.org
Established in 1945 to preserve peace through international cooperation and collective security. Composed of six main organs: General Assembly, the Security Council, the Economic and Social Council, the Trusteeship Council, the Secretariat, and the International Court of Justice.

Web Sites

MapMachine, National Geographic
http://plasma.nationalgeographic.com/mapmachine/index.html
Browse and search the world by name of. Includes a variety of maps.

Peace Corps
www.peacecorps.gov/wws/students/index.html
Learn about life in other countries.

The United States Commercial Service
www.export.gov/comm_svc/
Provides market research for many countries including trade shows, exporting information, and commercial business guides.

Subject Headings

- Americans—foreign countries
- cabinet officers—registers
- diplomatic and consular service
- gazetteers
- geography—encyclopedias
- government
- heads of state—registers
- international travel regulations
- political leadership—encyclopedias
- population—statistics—periodicals
- travel—safety measures—government policy—United States
- travel restrictions—government policy—United States

Cover Letters and Resumes: A Research Guide

Many times your cover letter and resume open the door to an interview. Cover letters must be crafted so that they spark an interest in the person who reads it—among all of the others that they receive. The cover letter should address the needs of the company and identify some of the corresponding talents of the applicant. The resources listed below will help get you started in creating an outstanding cover letter and resume. The resources listed below are available in the library and online. The terms and phrases listed in the subject headings below can be used to search for more materials in the library's catalog and research databases. If you need further assistance, please ask a librarian.

Writing Cover Letters

175 High-Impact Cover Letters, 3rd ed. by Richard H. Beatty. New York: John Wiley & Sons, 2002. ISBN: 0471210846.
Includes step-by-step instructions and many samples. Covers the five different types of cover letters: employer broadcast letters, search firm broadcast letters, advertising response letters, networking cover letters, and resume letters.

The Complete Idiot's Guide to the Perfect Resume by Susan Ireland. Indianapolis: Alpha Books, 2003. ISBN: 0028644409.
Offers strategy for your job hunt and how to polish and refine a resume.

Cover Letter Guide
www.susanireland.com/coverletterwork.htm
Brief instructional on writing cover letters with sample letters for you to use.

Cover Letters, Etc., Monster Resume Center
http://resume.monster.com/archives/coverletter/
Offers articles, tips, and information on all aspects of writing cover letters.

Cover Letters that Knock 'Em Dead, 5th ed. by Martin John Yate. Boston: Adams Media, 2002. ISBN: 1580627935.
Shows you how to create a compelling cover letter that will land you the interviews that you want.

The First and Best Cover Letters on the Internet, CareerLab
www.careerlab.com/letters/default.htm
Features 200 free cover letters to use in your job hunt.

Writing Resumes

10 Minute Resume
www.10minuteresume.com
Free service to write and publish a resume including posting to the Web, faxing your resume, and printing a copy.

The Complete Idiot's Guide to the Perfect Resume by Susan Ireland. Indianapolis: Alpha Books, 2003. ISBN: 0028644409.
Learn how to package yourself to potential employers by finding your strengths and selling points. Lots of sidebars, tips, and hits.

High Impact Resumes and Letters: How to Communicate Your Qualifications to Employers (High Impact Resumes and Letters), 8th ed. by Ronald L. Krannich and William J. Banis. Manassas, VA: Impact, 2002. ISBN: 1570231893.
This unique book provides step-by-step guidance on how to develop a personal profile that clearly communicates patterns of performance and accomplishments to employers.

Monster Resume Center
http://resume.monster.com
Offers resume tips by industry, dos, don'ts, and dilemmas.

The Overnight Résumé, 2nd ed. by Donald Asher. Berkeley, CA: Ten Speed Press, 1999. ISBN: 1580080413.
Basic guide to resume writing.

Resume Catalog: 200 Damn Good Examples, reprint ed. by Yana Parker. 1988. Berkeley, CA: Ten Speed Press, 1996. ISBN: 0898158915.
Includes sample resumes for adaptation or inspiration. Cover all levels of experience and a wide range of careers.

Resume Magic: Trade Secrets of a Professional Resume Writer, 2nd ed. by Susan Britton Whitcomb. Indianapolis: JIST Works, 2003. ISBN: 1563708914.
This guide gives techniques to improve resumes. Includes "before and after" samples.

Resumes in Cyberspace: Your Complete Guide to a Computerized Job Search, 2nd ed. by Pat Criscito. Hauppauge, NY: Barron's, 2000. ISBN: 0764114891.
Covers the different types of electronic résumés: scannable, e-mailable, and Web-formatted resumes.

Resumes that Knock 'Em Dead, 5th ed. by Martin John Yate. Boston: Adams Media, 2002. ISBN: 1580627943.
Shows you how to create a compelling resume that will land you the interviews that you want.

SUBJECT HEADINGS

- cover letters
- job applications
- resumes (employment)

Criminal Justice: A Research Guide

Students looking for a wide variety of criminal justice topics such as the court system, statistics about crimes, terrorism, technology and careers in criminal justice will find the resources listed below useful to start their research. Informational sources include books, journals, and online resources such as the library's databases and Web sites. Explore general topics in the encyclopedias and once you have narrowed your topic search for journal articles for supporting information. The terms and phrases listed in the subject headings below can be used to search for more materials in the library's catalog and research databases. If you need further assistance, please ask a librarian.

Sources of Definitions

Black's Law Dictionary, 7th ed. edited by Bryan A. Garner and Henry Campbell Black. St. Paul: West Group, 1999. ISBN: 0314241302.
Definitions of the terms and phrases of American and English jurisprudence, ancient and modern. Includes multiple meanings and examples of word usages with references to cases and laws.

Dictionary of Criminal Justice by George E. Rush. Guilford, CT: Dushkin Publishing Group, 2000. ISBN: 0070307091.
Dictionary of terms on law enforcement, courts, probation, parole, corrections, and related subjects. This dictionary focuses on terms commonly used in criminal justice practice.

National Criminal Justice Thesaurus. U.S. Department of Justice, National Institute of Justice, National Criminal Justice Reference Service. ISSN: 01986546.
Contains terms used to index literature in the NCJRS database. Source for finding synonyms and broader, narrower and related terms in the criminal justice field.

Guides to the Literature

Criminal Justice Information: How to Find It, How to Use It by Dennis C. Benamati, et al. Phoenix: Oryx Press, 1998. ISBN: 089774957X.
Advice on how to do research, establish connections, find information in the library, locate government and international information, as well as Web sources of interest.

Criminal Justice Research Sources, 4th ed. by Quint Thurman, Lee E. Parker, and Robert L. O'Block. Cincinnati: Anderson, 2000. ISBN: 0870848607.
Overview of criminal justice research sources with chapters on the research process, collecting data, indexing services, computerized literature searches, reference sources, NCJRS, statistical research, legal research, and Internet research.

Encyclopedias

American Justice edited by Joseph M. Bessette. 3 vols. Englewood Cliffs, NJ: Salem Press, 1996. ISBN: 0893567612.
Examine the legal system in the United States including the civil and criminal court systems, issues of constitutional law, legal history, and social justice.

Encyclopedia of Crime and Justice, 2nd ed. edited by Joshua Dressler. 4 vols. New York: Macmillan Reference USA, 2002. ISBN: 002865319X.
Includes information on the causes of crime, criminal behavior, crime prevention, institutions of criminal justice.

Encyclopedia of Criminology and Deviant Behavior edited by Clifton D. Bryant. 4 vols. Philadelphia: Brunner/Routledge, 2001. ISBN: 1560327723.
Covers topics related to criminology, deviant behavior, and other unusual sociological phenomena. Includes definition of each term, concise practical information, tables and diagrams, and a bibliography.

Encyclopedia of Violence, Peace and Conflict edited by Lester Kurtz. 3 vols. San Diego: Academic Press, 1999. ISBN: 012227010X.
Includes studies on "violence, peace, and conflict" from various points of view including criminology, cultural studies, economic, historical, political, biomedical, psychological, sociological, and public policy.

Violence in America edited by Ronald Gottesman and Richard M. Brown. 3 vols. New York: Charles Scribner's Sons, 1999. ISBN: 0684804875.
Covers many aspects of violence in the United States—physical, emotional, and psychological, as well as its impact on the environment and property.

Statistics

Crime State Rankings edited by Kathleen O. Morgan and Scott Morgan edited by Kathleen O. Morgan. Lawrence, KS: Morgan Quitno, 1994– . Annual. ISSN: 10774408.
Offers tables of crime data for U.S. states. Number of crimes, crime rates, and percent change are shown.

Sourcebook of Criminal Justice Statistics by Michael J. Hindelang. Washington, DC: Government Printing Office, 1973– . Annual. ISSN: 03603431.
Tables of data on criminal justice and related topics collected from governmental and private agencies. Also available in electronic version and on the Internet at *www.albany.edu/sourcebook/*.

Uniform Crime Reports, 1930–1997 (1998 to present available *www.fbi.gov/ucr/ucr.htm*). Washington, DC: Government Printing Office. ISSN: 00827592.
National crime statistics for use in law enforcement administration, operation, and management, including offenses cleared, persons arrested, homicide patterns, and law enforcement personnel. Also available on the Internet at *www.fbi.gov/ucr/ucr.htm.*

Web Sites

NCJRS (National Criminal Justice Reference Service) *Abstracts* Database
www.ncjrs.org/search.html
Contains documents, citations, and abstracts for journal articles, books, reports, hearings, studies, bibliographies, and statistical reports.

United Nations Crime and Justice Information Network
www.uncjin.org
Includes U.N. criminal justice country profiles, U.S. Department of Justice sources, recent U.S. Supreme Court decisions, the "World Criminal Justice Library Network," foreign and international law, criminological institutes, and libraries.

Subject Headings

- corrections—United States—periodicals
- crime—government policy—United States
- crime—research—United States—statistics
- crime—United States
- criminal justice—administration of—United States
- criminal justice—administration of—vocational guidance—United States
- criminal justice—administration of—United States—periodicals
- criminal statistics—United States—atlases
- law enforcement—vocational guidance—United States
- police—vocational guidance—United States
- prisoners—United States—statistics—periodicals

Current Events: A Research Guide

So many things can be learned by reading the day's news or listening to a broadcast. The daily news often becomes the topic for further discussions throughout the day. Online sources of information remain the most current; however, more in-depth coverage may follow in print newspapers. Some Web sites allow you to set up profiles so that news updates can be sent via e-mail, which gives viewers a personalized view of the news. This guide introduces the researcher to some of the basic informational sources on the topic. If you need further assistance, please ask a librarian.

General News Media

ABCNews.com: *http://abcnews.go.com*
The Chicago Tribune: *www.chicagotribune.com*
The Christian Science Monitor: *www.csmonitor.com*
CNN.com: *www.cnn.com*
The Dallas Morning News: *www.dallasnews.com*
The Houston Chronicle: *www.chron.com*
The Los Angeles Times: *www.latimes.com*
The New York Daily News: *www.nydailynews.com*
The New York Times: *www.nytimes.com*
The Philadelphia Inquirer: *www.philly.com/mld/inquirer/*
USA Today: *www.usatoday.com*
The Wall Street Journal: *www.wsj.com*
The Washington Post: *www.washingtonpost.com*

Financial Media

Bloomberg: *www.bloomberg.com*
CNN Money: *http://money.cnn.com*

Web Sites

1st Headlines
http://1stheadlines.com
News headlines from around the world.

Documents in the News, University of Michigan
www.lib.umich.edu/govdocs/docnews.html
Features links to government publications that have been mentioned in the news.

HeadlineSpot
www.headlinespot.com
Sources for 56 U.S. metropolitan areas, 50 states, 59 countries, 27 industries and dozens of subjects from arts and automobiles to technology and travel.

Newshour Extra with Jim Lehrer
www.pbs.org/newshour/extra/
A news site especially for students. Produced by PBS.

Newspage
www.individual.com
Create customized news profile by selecting topics or companies.

Online Newspapers from Around the World
www.ipl.org/div/news/
Links to newspapers. Organized by region and country.

Yahoo! News
http://news.yahoo.com
Lists top headlines from major news sites.

MAGAZINES

Foreign Affairs Magazine
www.foreignaffairs.org
Covers international affairs and foreign policy. Published since 1922. This site features a searchable archives dating back to 1979.

MSNBC Newsweek Frontpage
www.newsweek.com
Online version of the print magazine.

Time Online Edition
www.time.com
Includes specials such as *Time's* list of most influential people.

U.S. News & World Report
www.usnews.com
World news as reported.

INTERNATIONAL NEWS

BBC World Service
www.bbc.co.uk/worldservice/index.shtml
The British Broadcasting Corporation (BBC) offers an array of broadcasting options for world news.

World Press Review
www.worldpress.org
Provides view of the political and economic climate outside of the United States through translating, reprinting, analyzing, and contextualizing the best of the international press from more than 20 languages.

Decades: A Research Guide

It is often the daily activities versus the major events in history that go unrecorded. Many of those very fascinating and interesting facts can be found in the many resources below that tell just how life was back in the olden days. Chronologies often place people and events in perspective; however, it is the personal narrative that often tells more of the story. Resources can be found in the library and on the Internet to assist visitors in beginning their research. The terms and phrases listed in the subject headings below can be used to search for more materials in the library's catalog and research databases. If you need further assistance, please ask a librarian.

Books

American Century by Harold Evans, Gail Buckland, and Kevin Baker. New York: Knopf, 1998. ISBN: 0679410708.
Short articles, photos and sidebars give perspective on the most significant personalities and political episodes of the century.

American Chronicles: Year by Year Through the Twentieth Century by Lois G. Gordon and Alan Gordon. New Haven, CT: Yale University Press, 1999. ISBN: 0300075871.
Chapters for each include major events in the news and selections of quotations and ads.

American Decades edited by Vincent Tompkins, et al. 10 vols. Detroit: Gale Research, 1994–2001. ISBN: 0810357224.
Each volume focuses on a particular decade (1900–1909, 1910–1919, etc.) and begins with a month by month chronology of events both within the United States and in the world at large. The remainder of the book is divided into broad categories like business, the arts, law, religion, sports, etc.

The American Years: A Chronology of United States History by Ernie Gross. New York: Charles Scribner's Sons/Macmillan Library Reference, 1999. ISBN: 0684805901.
General chronology beginning with 1776 and ending in the year 1997.

Chronicle of America by Clifton Daniel and John W. Kirshon. New York: Dorling Kindersley, 1997. ISBN: 0789420910.
Year by year compendium of key news stories and major events. This could be used to find events from specific years as well as specific decades. There are many illustrations, and the articles are easy to read.

Day by Day series published by Facts On File . Provides a day by day chronology of each day of a particular decade.

- *Day by Day: The Eighties* by Ellen Meltzer and Marc Aronson. 2 vols. New York: Facts On File, 1995. ISBN: 0816015929.
- *Day by Day: The Fifties* by Jeffrey D. Merritt and Steven L. Goulden. New York: Facts On File, 1979. ISBN: 0871963833.
- *Day by Day: The Forties* by Thomas M. Leonard, Richard D. Burbank, and Steven L. Goulden. New York: Facts On File, 1977. ISBN: 0871963752.
- *Day by Day: The Seventies* by Thomas M. Leonard, Cynthia Crippenk, and Marc Aronson. 2 vols. New York: Facts On File, 1996. ISBN: 081601020X.
- *Day by Day: The Sixties* by Thomas Parker and Douglas Nelson. 2 vols. New York: Facts On File, 1983. ISBN: 0871966484.

DK Visual Timeline of the Twentieth Century edited by Simon Adams. New York: Dorling Kindersley, 1996. ISBN: 0789409976.
Covers major events and discoveries of the 20th century in chronological sequence.

Historic Events for Students: The Great Depression by Richard Clay Hanes and Sharon M. Hanes. 3 vols. Detroit: Gale Group, 2002. ISBN: 0787657018.
Covers all aspects of the era in great detail: economic and political, as well as the New Deal.

Twentieth Century America: A Primary Source Collection from the Associated Press. 10 vols. Danbury, CT: Grolier Educational, 1995. ISBN: 0717274942.
Twentieth-century chronology told by excerpts from Associated Press news stories.

U.S.A. Sixties. 6 vols. Danbury, CT: Grolier Educational, 2001. ISBN: 0717295036.
Encyclopedia about the 1960s. The entries are arranged alphabetically. Illustrated with photos, this work covers all aspects of the 1960s: political and cultural, as well as the arts.

Web Sites

American Cultural History: The Twentieth Century
http://kclibrary.nhmccd.edu/decades.html
A series of Web guides on the decades of the 20th century. The pages are being prepared by the reference librarians at Kingwood College.

The Decades Project
www.ptsd.wednet.edu/district/resources/bookmarks/cn/centurynews.html
An index of Web sites arranged by decades from 1900–1990.

Infoplease: Year by Year
www.infoplease.com/yearbyyear.html
Lists notable events for each year and decade including U.S. and world history and events in economics, sports, entertainment, and science. Listings also include brief biographies of people who died during the decade.

Subject Headings

- nineteen forties—chronology
- nineteen fifties—chronology (etc.)
- United States—history—20th century
- United States—history—20th century—sources
- United States—history—chronology
- United States—politics and government—20th century

Decorative Arts Identification: A Research Guide

Research on objects belonging to one of the categories listed below involves the consultation of specific handbooks and guides. The following decorative arts reference books are listed as sources that may help in identifying the origins of various objects. These books are also useful guides to visual images, history, terminology, and bibliography for further research. This guide introduces researchers to some of the basic informational sources on the topic. The terms and phrases listed in the subject headings below can be used to search for more materials in the library's catalog and research databases. If you need further assistance, please ask a librarian.

Furniture

The Complete Guide to Furniture Styles by Louise Ade Boger. New York: Waveland Press, 1997. ISBN: 0881339393.
Broad survey history organized by historical period with subdivisions for geographic coverage and specific styles: e.g., Louis XIV, Empire, French Provincial, Chippendale, American Colonial, and Federal. Text followed by black-and-white plates of over 500 objects. Includes bibliography.

The Encyclopedia of Furniture, 3rd ed. by Joseph Aronson. New York: Crown, 1965. ISBN: 0517037351.
Short entries on world furniture, from antiquity to the early modernist period. Also includes longer essays on major countries, types of furniture, and styles.

Fake, Fraud or Genuine? Identifying Authentic American Furniture by Myrna Kaye. Boston: Little, Brown, 1991. ISBN: 0821218255.
Sourcebook of identification techniques, problems in construction, inspection processes, and various clues to spotting fraudulently identified pieces. Covers 17th–19th-century furniture. Many photographs, often of test cases and their details.

Field Guide to American Antique Furniture by Joseph T. Butler, Kathleen Eagen Johnson, and Ray Skibinski. New York: Facts On File, 1985. ISBN: 0816010080.
Traces the history of American furniture from 17th through early 20th century. 1700 illustrations—line drawings arranged in chronological sequence by type of furniture.

Glass and Stained Glass

An Illustrated Dictionary of Glass by H. Newman. London: Thames & Hudson, 1977. ISBN: 0500232628.
Contains entries for terms about wares, materials, processes, forms and decorative styles, and entries on principal glassmakers, decorators, and designers, from antiquity to the present.

Our House Antiques—Confusing Patterns
www.ourhouseantiques.com/pattid.htm
Identifies some patterns of glass that are commonly mistaken for one another or which are frequently misidentified as to manufacturer or pattern name.

Stained Glass by Lawrence Lee, George Seddon, and Francis Stephens. New York: Crown, 1976. ISBN: 0517527286.
Standard book on stained glass. Includes illustrations and bibliography.

Silver

American Silver A History of Style, 1650–1900 by Graham Hood. New York: Praeger, 1971.
Traces development of American silver styles. Includes 286 photographs and bibliography.

The Book of Old Silver, English, American, Foreign by Seymour B. Wyler. New York: Crown, 1937. ISBN: 051700089X.
Essays on historical aspects of old silver, e.g., laws, frauds, collection and care, types of silver items, Sheffield plate. American and European countries' hallmarks. Index to marks.

An Illustrated Dictionary of Silverware by Harold Newman. London: Thames & Hudson, 1987. ISBN: 0500281963.
Contains 2,373 entries on British and North American silver. Techniques, styles, leading designers, and makers from circa 1500 to the present; emphasis on 17th–19th centuries.

Silver Mark Finder
www.instappraisal.com
Indentifies the manufacturer, years and country of silver marks.

POTTERY AND PORCELAIN

American Ceramics 1876 to the Present by Garth Clark. New York: Abbeville Press, 1990. ISBN: 0896597431.
Magnificently illustrated survey decade by decade, contains chronologies, selected exhibitions, biographies, bibliography.

Encyclopaedia of British Pottery and Porcelain Marks by Geoffrey A. Godden, London: Barrie & Jenkins, 1991. ISBN: 0257657827.
Alphabetical listings of potteries or firms, short biographical data, dates if known, cross-references and reproduced marks if known.

An Illustrated Dictionary of Ceramics by George Savage and Harold Newman. London: Thames & Hudson, 2000. ISBN: 0500273804.
Alphabetical listing of 3054 terms on wares, materials, processes, styles, patterns, and shapes from antiquity to the present day. Many black-and-white illustrations.

Kovels' Dictionary of Marks: Pottery and Porcelain: 1850–Present by Ralph M. Kovel and Terry H. Kovel. New York: Random House, 1996. ISBN: 0517559145.
Lists of marks. Vocabulary of marks, dating systems used by specific factories.

Porcelain Marks
http://users.skynet.be/rutrene/marks.html
Includes information about the marks on European porcelain.

SUBJECT HEADINGS

- art objects—dictionaries
- collectors and collecting
- decorative arts—dictionaries
- furniture—collectors and collecting—United States
- furniture—dictionaries
- furniture—expertising—United States
- glass—dictionaries
- porcelain—dictionaries
- porcelain—marks
- pottery—dictionaries
- pottery—marks
- silverwork—dictionaries
- silverwork—United States—history

Demographics: A Research Guide

Demographics are the characteristics of the population that influence consumption of products and services. Information to create demographic profiles originate from a variety of sources, both private and governmental. Characteristics include population trends, age, gender, race or ethnicity, education, income, crime rates, voting statistics, and occupations. Sources of information can be found readily in print and online sources. Researchers use this information in order to understand market trends to introduce new products or businesses to a group or geographic area. The terms and phrases listed in the subject headings below can be used to search for more materials in the library's catalog and research databases. If you need further assistance, please ask a librarian.

Reference Books

Editor & Publisher Market Guide. New York: Editor & Publisher, 1924– . Annual. ISSN: 10820779.
Business and demographic data on 1,600 U.S. daily newspaper cities. Includes spending data and demographic trends.

Rand McNally Commercial Atlas & Marketing Guide. Chicago: Rand McNally, 1983– . Annual.
This atlas provides current information on demographics for marketing. Maps and tables show populations, income, and housing and sales data (with projections) for the United States.

Sourcebook of ZIP Code Demographics, 17th ed. Vienna, VA: ESRI Press, 2003. Annual. ISBN: 1589480570.
Provides detailed information about every ZIP Code and county in the United States. Contains more than 70 demographic variables including Census 2000 information and updated demographics, spending potential data, and business data for every United States ZIP Code and county.

Reports and Data Sources

American Community Survey
www.census.gov/acs/www/
Demographic, social, economic, and housing data for over 800 geographical areas in narrative and tabular formats.

American Factfinder
http://factfinder.census.gov
U.S. Department of Commerce statistics on population, housing, industry, and business. Includes the entire 2000 Census of Population and Housing.

Ameristat
www.ameristat.org
Provides summaries—in graphics and text—of the demographic characteristics of the U.S. population.

Ersys
www.ersys.com
Statistical information on American cities and metropolitan areas in all fifty states, there is information on education, transportation, ethnicity, and more.

The International Data Base (IDB)
www.census.gov/ipc/www/idbnew.html
Computerized data bank that contains statistical tables of demographic and socioeconomic data for 227 countries and areas of the world.

School District Demographics
http://nces.ed.gov/surveys/sdds/index.asp

Provides access to school district geographic and demographic data useful for describing and analyzing characteristics of school districts, children, and K–12 education.

Journals

American Demographics. American Demographics; Dow Jones. Stamford, CT: Cowles Business Media, 1979– . Monthly.

Covers American demographics and consumer trends.

Associations

American Marketing Association
www.marketingpower.com

With over 38,000 members worldwide, AMA is the largest professional association for marketers. Provides relevant marketing information, journal publications, research, case studies, and the best practices in marketing.

Subject Headings

- brand choice—United States
- consumers—United States
- consumers' preferences—United States
- income—United States—statistics
- market surveys—United States
- purchasing power—United States—statistics
- United States—commerce—statistics
- United States—population—statistics

Diseases: A Research Guide

When disease or the possibility of disease strikes you or someone close to you, a natural instinct is to learn as much as possible about the disease in order to understand its impact on the body and to enable the patient and doctor to have more engaging discussions about it. This guide introduces the researcher to some of the basic informational sources on the topic. The terms and phrases listed in the subject headings below can be used to search for more materials in the library's catalog and research databases. If you need further assistance, please ask a librarian.

Reference Books

The Merck Manual of Medical Information, 2nd ed., Home ed. by Mark H. Beers. Simon & Schuster, 2003. ISBN: 0911910352. Also available online at *www.merck.com/pubs/.*
Comprehensive look at many disorders, symptoms and treatments. Based on the authoritative textbook of medicine, *The Merck Manual.*

Mosby's Handbook of Diseases edited by Rae Langford and June M. Thompson. St. Louis: Mosby, 2000. ISBN: 032300895X.
Profile of 231 adult and pediatric diseases and conditions including definitions and descriptions, complications, frequently ordered diagnostic tests, and treatments.

Mosby's Pediatric Nursing Reference by Linda A. Sowden and Cecily Lynn Betz. St. Louis: Mosby, 2000. ISBN: 0323009352.
Includes coverage of major pediatric medical and surgical conditions, diagnostic tests, and procedures in a convenient alphabetical format.

Professional Guide to Diseases, 7th ed. Springhouse, PA: Lippincott Williams & Wilkins, 2001. ISBN: 1582550735.
Information on diseases, findings, and treatments for over 600 diseases. Organized by body system. Appendices cover rare diseases, cultural considerations, and community resources.

Web Sites

AIDS Gateway to the Internet
www.aids.org
Provides fact sheets on AIDS as well as prevention, treatment, and current news.

Hepatitis.org
www.hepatitis.org
Provides information on the various types of hepatitis including alcoholic hepatitis, hepatitis in the workplace, viral hepatitis, hepatitis C, and more.

MayoClinic.com
www.mayoclinic.com
MayoClinic.com's medical experts and editorial professionals bring you information on diseases, conditions, treatment, and prevention.

Medline Plus
www.nlm.nih.gov/medlineplus/healthtopics.html
Offers a variety of information on diseases, disorders, and conditions including encyclopedia entries and related organizations.

The Merck Manual of Medical Information, 2nd ed., Home ed.
www.merck.com/pubs/
Comprehensive look at many disorders, symptoms and treatments. Based on the authoritative textbook of medicine, *The Merck Manual.*

WebMD, Diseases and Conditions
http://my.webmd.com/medical_information/condition_centers/default.htm
Guide to the basics, treatment, prevention, community, experts, and the latest news.

Organizations

American Cancer Society
www.cancer.org
Publishes information to support cancer patients and survivors including early detection, treatment options, clinical trials. Provides access to local information.

American Diabetes Association
www.diabetes.org
Offers information about diabetes prevention, weight loss, nutrition, recipes, and more.

Centers for Disease Control and Prevention
www.cdc.gov
Includes information about health and safety topics and access to publications, data, and statistics.

National Institute of Arthritis and Musculoskeletal and Skin Diseases Information Clearinghouse
www.niams.nih.gov
Health information includes a list of the most frequently asked questions about various types of arthritis and musculoskeletal and skin diseases and their answers. Also includes related organizations.

National Institute of Diabetes and Digestive and Kidney Diseases
www.niddk.nih.gov
Conducts biomedical research, disseminating research findings and health information to the public. Offers information about diabetes, digestive endocrine and metabolic diseases, hematologic diseases, kidney and urologic disease, nutrition and weight loss, and statistics.

National Institute of Health, U.S. Department of Health and Human Services
www.nih.gov
Part of the U.S. Department of Health and Human Services. Primary Federal agency for conducting and supporting medical research to improve people's health and save lives.

Office of Rare Diseases, National Institute of Health
http://rarediseases.info.nih.gov
Genetic and rare disease information center for major topics of interest in rare diseases research.

World Health Organization
www.who.int/en/
Provides access to research database including all WHO publications from 1948 onwards and articles from WHO-produced journals and technical documents from 1985 to the present.

Subject Headings

- communicable diseases
- disease—handbooks
- disease—nurses' instruction
- medicine, popular—handbooks, manuals, etc.
- symptoms—handbooks, manuals, etc.
- vaccination—statistics and numerical data—handbooks

Economics: A Research Guide

Economics is the study of how human beings allocate resources to produce various commodities and how those commodities are distributed for consumption among people in society. Researchers will find a variety of resources at their fingertips and through text, data sets, charts, and graphs will explain and illustrate the theories that lay the foundation for the study of economics. This guide introduces the researcher to some of the basic informational sources on the topic. The terms and phrases listed in the subject headings below can be used to search for more materials in the library's catalog and research databases. If you need further assistance, please ask a librarian.

General Reference Resources

Encyclopedia of American Economic History: Studies of the Principal Movements and Ideas by Glenn Porter. 3 vols. New York: Charles Scribner's Sons, 1980. ISBN: 0684162717.
Lists comprehensive articles by the scholars in the field of economics.

Encyclopedia of the New Economy
http://hotwired.wired.com/special/ene/
Defines terms related to the new economy in which information technology plays a significant role and enables producers of both the tangible and intangible to compete efficiently in global markets.

Gale Encyclopedia of U.S. Economic History edited by Thomas Carson and Mary Bonk. 2 vols. Detroit: Gale Group, 1999. ISBN: 0787638889.
Includes 1,000 entries that range from one paragraph to several pages in length covering era overviews, event/movement profiles, biographies, business/industry profiles, geographic profiles, and more.

The New Palgrave: A Dictionary of Economics by John Eatwell, Murray Milgate, Peter K. Newman, et al. 4 vols. New York: Macmillan, 1987. ISBN: 0333372352.
Four volumes in length, its entries range from a page or so up to ten pages. Bibliographies are included.

The New Palgrave Dictionary of Economics and the Law by Peter K. Newman. 3 vols. New York: Macmillan Reference, 1998. ISBN: 1561592153.
Includes 399 signed articles written by American or British academics from major institutions. Entries are interdisciplinary and encyclopedic in range and depth. Covers topics like the functioning of markets, property rights and their enforcement, and financial obligations.

The New Palgrave Dictionary of Money and Finance by Peter K. Newman, Murray Milgate, and John Eatwell. 3 vols. New York: Macmillan, 1992. ISBN: 156159041X.
Contains over 1,000 essays by leading professionals covering finances, and economics. Includes over 200 tables and diagrams and bibliographies for all entries.

The Oxford Encyclopedia of Economic History edited by Joel Mokyr. 5 vols. New York: Oxford University Press, 2003. ISBN: 0195105079.
Includes more than 875 articles written and signed by international contributors. Organized by chronological and geographic boundaries and related subfields such as agricultural history, demographic history, business history, and the histories of technology, migration, and transportation. Volume topic includes: vol. 1: accounting and bookkeeping; contract labor and the indenture system; vol. 2: cooperative agriculture and farmer cooperatives; Jonathan Hughes; vol. 3: human capital; Mongolia; vol. 4: Monte di Piet; Spain; vol. 5: spices and spice trade; zoos and other animal parks.

Survey of Social Science. Economics Series edited by Frank Northen Magill and Demos Vardiabasis. 5 vols. Pasadena, CA: Salem Press, 1991. ISBN: 0893567256.
393 articles focusing on 12 key areas of economics. Includes topics such as monetary theory, micro- and macroeconomics, banking and fiscal policy, as well as health care, housing, poverty, and social security issues.

Journals

American Economic Review
Contemporary Economic Policy
Economic Journal
Economist
Journal of Economic History
Journal of International Economics
OECD Observer
Quarterly Journal of Economics

Statistics

Bureau of Economic Analysis
www.bea.doc.gov
Provides timely, relevant, and official economic indicators including industry data such as gross domestic product, international trade and investment data regional estimates including state personal income, gross state product, and regional multipliers.

Bureau of Labor Statistics
www.bls.gov
Provides data on wages, unemployment, occupation, and prices data.

Economagic.com
www.economagic.com
Economagic.com gives you easy access to more than 100,000 data series including state, metro and county employment data compiled by federal statistical agencies. The site will create spreadsheet files of data online, as well as graphing data in your Internet browser. Registered users can generate forecasts from historical data.

Economic Census Reports, Bureau of the Census
www.census.gov
Provides a detailed portrait of the Nation's economy once every 5 years, from the national to the local level.

FedStats, Federal Interagency Council on Statistical Policy
www.fedstats.gov
Provides state and local data profiles drawn from multiple federal statistical agencies.

Regional Economic Conditions, Federal Deposit Insurance Corporation
www2.fdic.gov/recon/
Provides access to state, MSA, or county data. View standard graphs, tables, and maps depicting economic conditions, and how they have changed over time.

Subject Headings

- economic history—encyclopedias
- economics—dictionaries
- law—economic aspects—dictionaries
- law and economics—dictionaries
- law and economics—encyclopedias
- United States—economic conditions

Education: A Research Guide

Students seeking information on the latest teaching techniques or support of curriculum development will find a great starting point for their research using the materials listed below. Information such as educational concepts, special education, trends, and issues in education are readily available to educators and students. The terms and phrases listed in the subject headings below can be used to search for more materials in the library's catalog and research databases. If you need further assistance, please ask a librarian.

Reference Books

Biographical Dictionary of Modern American Educators edited by Frederik Ohles, Shirley M. Ohles, and John G. Ramsay. Westport, CT: Greenwood Press, 1997. ISBN: 0313291330.
Biographical sketches are listed in alphabetical order by last names of subjects. Includes a list of acronyms/abbreviations, a combined name and subject index, and appendixes.

The Cyclopedic Education Dictionary edited by Carol S. Spafford, Augustus J. Itzo Pesce, and George S. Grosser. Albany, NY: Delmar, 1998. ISBN: 0827384750.
A concise dictionary that provides researchers with valuable educational terms, concepts, issues, and strategies.

The Dictionary of Educational Terms edited by David Blake and Vincent Hanley. Brookfield, VT: Arena, 1995. ISBN: 1857422570.
Arranged alphabetically, this dictionary contains entries covering basic educational terms. Included also is a useful list of acronyms.

Education, A Guide to Reference and Information Sources by Nancy P. O'Brien and Lois Buttlar. Englewood, CO: Libraries Unlimited, 2000. ISBN: 1563086263.
Fourteen chapters, each based on a broad subject category, such as educational administration and management, educational psychology, and special education, cover various types of reference materials and journals.

Education Index. New York: H. W. Wilson, 1929– . Annual with monthly updates. ISSN: 00131385.
Primarily a periodical index, it draws from more than 400 English language journals, including some not indexed elsewhere. Following the main body is a separate listing of book reviews.

Educational Media and Technology Yearbook. Littleton, CO: Libraries Unlimited, 1985– . Annual. ISSN: 87552094.
Focuses on trends and issues; school library and media, North American organizations and associations, graduate programs, and mediagraphy—print and nonprint sources.

Encyclopedia of American Education, 2nd ed. edited by Harlow G. Unger. 3 vols. New York: Facts On File, 2001. ISBN: 0816043442.
Illustrated with many photos, all entries include at least one bibliographic reference.

The Encyclopedia of Education, 2nd ed. edited by James W. Guthrie. 8 vols. New York: Macmillan Reference, 2003. ISBN: 002865594X.
Focus is on U.S. practices and institutions, from preschool to higher education.

Encyclopedia of Language and Education edited by David Corson. 8 vols. Boston: Kluwer, 1997. ISBN: 0792345959.
Covers a large range of topics in language and education. Each volume is dedicated to a specific topic and contains in-depth signed articles written by eminent scholars in the field.

Handbook of Research on Teaching, 4th ed. edited by Virginia Richardson. Washington, DC: American Educational Research Association, 2001. ISBN: 0935302263
Consists of signed articles. Includes bibliographical references and indexes.

Philosophy of Education: An Encyclopedia edited by Joseph J. Chambliss. New York: Garland, 1996. ISBN: 081531177X.
Contains signed articles with bibliographic references covering people and concepts from Plato to Montessori.

Who's Who in American Education. Owings Mills, MD: National Reference Institute, 1988– . Annual publication. ISSN: 10467203.
Brief biographies are arranged alphabetically by last names. Includes a table of acronyms/abbreviations, a "professional index" by academic discipline, and a "teaching awards" index.

World of Learning. London: Allen & Unwin, 1947– . Annual. ISSN: 00842117.
Source of names, addresses, and other information about universities and other academic, artistic, or scholarly organizations around the world.

Web Sites

The Educator's Reference Desk
www.eduref.org
Provides links to useful Web sites with information about citation styles, links to directories, government resources, law, intellectual freedom, libraries, mass media, school rankings, and statistics.

ERIC Database
www.eduref.org/Eric/
Provides access to resources in education through citations from 1966 through April 2004 and ERIC Journal citations from 1966 through March 2004. Full-text materials can be obtained through a fee-based service or from a library that maintains a collection of ERIC documents.

ERIC Digests
www.ericdigests.org
Contains short reports on topics of current interest in education.

National Center for Education Statistics
http://nces.ed.gov
Federal agency that collects and analyzes data related to education in the United States and other countries.

Subject Headings

- education
- educational technology
- instructional materials
- teaching materials

Elections: A Research Guide

Informed students, researchers, and citizens have a variety of resources from which to check on current and historical elections, candidates, platforms and issues. They can also check to answer frequently asked questions like where to register to vote and dates of the next election. This guide introduces the researcher to some of the basic informational sources on the topic. The terms and phrases listed in the subject headings below can be used to search for more materials in the library's catalog and research databases. If you need further assistance, please ask a librarian.

Reference Books

Elections A to Z by John L. Moore. Washington, DC: Congressional Quarterly, 1999. ISBN: 1568022077.
An illustrated encyclopedia covering every aspect of elections in the United States.

Guide to the 2004 Presidential Election by Michael L. Goldstein. Washington, DC: CQ Press, 2003. ISBN: 1568028482.
An objective journey through the presidential election process including biographical material on candidates, important issues, and developments.

The Routledge Historical Atlas of Presidential Elections by Yanek Mieczkowski and Mark C. Carnes. New York: Routledge, 2001. ISBN: 0415921333.
A summary of how people voted in presidential elections from 1788 to 2000.

Web Sites: Candidates and Issues

Democratic National Committee
www.democrats.org
Official Web site of the National Democratic Party. States all have their own committees.

League of Women Voters
www.lwv.org
A nonpartisan political organization that urges the informed and active input of citizens and works to increase understanding of major public policy issues.

National Republican Party
www.rnc.org
Official Web site of the National Republican Party. Each state has additional Web sites on a regional basis.

Politics
www.politics.com
Provides a list of all candidates from the 2000 election, including biographical information and links to available official and unofficial Web sites. Also includes information about the current administration and elections in other states.

Project Vote Smart
www.vote-smart.org
Nonpartisan source of information and analysis of candidates, issues, and ballot measures. Includes biographical information about office holders and candidates.

Election News and Information

C-SPAN
www.c-span.org
Provides coverage of politics and elections, including video footage of speeches and advertisements, a calendar of campaign events, and links to the Web sites of the major candidates.

Election Information, Office of the Clerk, U.S. House of Representatives
http://clerk.house.gov/members/electionInfo/index.html
Provides official federal election vote counts, as far back as 1920.

United States Election Project
http://elections.gmu.edu
Provides timely and accurate election statistics, electoral laws, research reports, and other useful information regarding the United States electoral system.

United States Elections 2004
http://usinfo.state.gov/products/pubs/election04/homepage.htm
Provides an introductory overview of the American electoral process.

U.S. Electoral College, National Archives and Records Administration
www.archives.gov/federal_register/electoral_college/
Explanation of the electoral college including statistics and historical accounts.

Election Laws and Finance

Federal Election Commission Financial Reports and Data
www.fec.gov/finance_reports.html
Find financial disclosure reports from congressional and presidential campaigns, including political parties and political action committees.

Opensecrets
www.opensecrets.org
Sponsored by the nonpartisan Center for Responsive Politics, this Web site provides a guide to campaign donations and the role of money in elections.

Election and Voting System Reform

Center for Voting and Democracy
www.fairvote.org
A nonpartisan organization that studies how voting systems affect participation, representation and governance.

National Commission on Federal Election Reform
www.reformelections.org
This nonpartisan group was organized to study the federal election system and to recommend ways to improve its fairness and accuracy.

Subject Headings

- elections—United States—encyclopedias
- political campaigns—United States—history
- political parties—United States—encyclopedias
- political parties—United States—platforms—history
- presidential candidates—United States—history
- presidents—United States—election

Elections for Kids: A Research Guide

Each year elections take place in every town and city throughout the United States. An election is the process by which people vote for candidates that they want to govern them. We not only vote so that people can lead our government, but we also vote on proposals or questions that may result in a new hospital or library, or changes to the city charter. The basis of democratic government is that citizens have the right to choose the officials who will govern them. Elections play a very important part in our daily lives. This guide introduces researchers to some of the basic informational sources on the topic. The terms and phrases listed in the subject headings below can be used to search for more materials in the library's catalog and research databases. If you need further assistance, please ask a librarian.

Books for Grades 3–5

Arnold for President by Craig Bartlett and Tom Parsons. New York: Simon Spotlight/Nickelodeon, 2000. ISBN: 068983361X.
When Arnold and Helga compete in a race for class president in fourth grade, Arnold learns just how a democracy works.

Class President by Johanna Hurwitz and Sheila Hamanaka. New York: Morrow Junior Books, 1990. ISBN: 0688091148.
Julio hides his own leadership ambitions in order to help another candidate win the nomination for class president.

Grace's Letter to Lincoln by Connie Roop, Peter Roop, and Stacey Schuett. New York: Hyperion Books, 1998. ISBN: 0786812966.
On the eve of the 1860 presidential election, young Grace decides to help Abraham Lincoln get elected by advising him to grow a beard.

Books for Grades 6 and Up

The Misfits by James Howe. New York: Atheneum Books for Young Readers, 2001. ISBN: 0689839553.
Four students who do not fit in at their small-town middle school create a third party for the student elections to represent students who have been called names.

Girl Reporter Rocks Polls! by Linda Ellerbee. New York: Avon Books/HarperCollins, 2000. ISBN: 0064407608.
School newspaper reporter Casey Smith uncovers a plot to sabotage the elections at her middle school.

The Kid Who Ran for President by Dan Gutman. New York: Scholastic, 1996. ISBN: 0590939882.
With his friend as campaign manager and his former babysitter as running mate, twelve-year-old Judson Moon sets out to become the President of the United States.

A Time for Courage: The Suffragette Diary of Kathleen Bowen by Kathryn Lasky. New York: Scholastic, 2002. ISBN: 0590511416.
A 1917 diary of thirteen-year-old Kathleen Bowen's life in Washington, DC, that reveals her concerns for the battle for women's suffrage, war in Europe, and her family.

Nonfiction

The Electoral College by Martha S. Hewson. Philadelphia: Chelsea House, 2002. ISBN: 0791067904.
Covers the 2000 election (Bush vs. Gore), the creation and original process of the electoral college, and how the electoral college works today.

The Fifteenth Amendment: African-American Men's Right to Vote by Susan Banfield. Springfield, NJ: Enslow, 1998. ISBN: 0766010333.
Examines the Amendment that gave African-American men the right to vote and discusses the struggle that took place to regain this right when it was denied.

The Nineteenth Amendment: Women's Right to Vote by Judy Monroe. Springfield, NJ: Enslow, 1998. ISBN: 0894909223.
The history of the women's suffrage movement in the United States that culminated with the passage of the constitutional amendment giving women the right to vote.

Presidential Elections and Other Cool Facts by Syl Sobel and Jill Wood. Hauppauge, NY: Barron's Educational Series, 2000. ISBN: 0764114387.
This book looks at the rules for presidential elections, the electoral college, a campaign, and the order of succession if something happens to the president.

Running for Public Office by Sarah De Capua. New York: Children's Press, 2002. ISBN: 051627368X.
Describes the process of running for a public office, including the planning and organizing of a campaign, the campaign trail, and the election day.

Web Sites

Congress for Kids: Elections
www.congressforkids.net/Elections_index.htm
Everything you need to know about elections from how the candidates are chosen, voting methods, and an explanation of the electoral college.

Project Vote Smart
www.vote-smart.org
Candidate backgrounds, issue positions, voting records, campaign finances, and performance evaluations.

The Library of Congress: Elections...the American Way
http://learning.loc.gov/learn/features/election/home.html
This site reviews the history of voting in America, with links to source documents such as political posters and recorded debates.

Take Your Kids to Vote
www.takeyourkidstovote.org/youth/index.htm
Activities for future voters of all ages.

The 30 Second Candidate
www.pbs.org/30secondcandidate/
An overview of political commercials organized on a time line.

Yahooligans the Big Picture: Elections
http://yahooligans.yahoo.com
Go to main Yahooligans page and search term "Elections" for a wealth of information on this topic.

Subject Headings

- African Americans—suffrage
- campaign management—United States
- electioneering—United States
- elections
- electoral college—United States
- presidents—election
- presidents—United States—election
- women—suffrage

Engineering: A Research Guide

Engineering originates from principles of science that are used to design structures and products of all kinds. The term *engineering* comes from the Latin word *ingeniare*, which means *to design* or *to create*. Information in the field of engineering can be found in handbooks, books, journals, conference proceedings, standards, technical reports, and Web resources. Additionally, professional engineering societies are a source of engineering publications. This guide introduces researchers to some of the basic informational sources on the topic. The terms and phrases listed in the subject headings below can be used to search for more materials in the library's catalog and research databases. If you need further assistance, please ask a librarian.

Reference Books

Handbook of Engineering Fundamentals, 4th ed. by Ovid W. Eshbach, et al. New York: Wiley, 1990. ISBN: 0471245534.
One-volume reference for engineers and students in engineering fields, especially mechanical engineering.

McGraw-Hill Encyclopedia of Science & Technology. New York: McGraw-Hill, 2002. ISBN: 0079136656.
No other single resource provides such comprehensive and authoritative coverage of all the major disciplines of science and technology.

Van Nostrand's Scientific Encyclopedia, 9th ed. by Glenn D. Considine and Peter H. Kulik. New York: Wiley-Interscience, 2002. ISBN: 0471332305.
This two-volume set ranges across all scientific disciplines, as well as many areas of engineering and technology.

Web Sites

EEVL Engineering Section
www.eevl.ac.uk/engineering/index.htm
Central access point to selected networked engineering information. EEVL is the Internet Guide to Engineering, Mathematics, and Computing.

Engineer on a Disk
http://claymore.engineer.gvsu.edu/eod/
Quick online tutorials in many areas of engineering.

Engineers Edge
www.engineersedge.com
Online calculators, equation charts, part specifications (e.g., drill sizes, gauge charts), etc.

Engineering.com
www.engineering.com
Resource tool for the global engineering community with a leading business-to-business Internet marketplace for engineering products and services.

FreeCalc.com: Online Application for Engineering Design
www.freecalc.com
Online applications for engineering design from Beacon Engineers.

Global Engineering Documents
http://global.ihs.com
From IHS Engineering, this Web site is a source for industry, government, and military standards.

Michigan eLibrary/Engineering
http://mel.lib.mi.us/viewtopic.jsp?id=1832&pathid=2373#M
Michigan eLibrary is an information gateway to selected Internet resources, full-text magazines, newspapers, electronic books, online practice tests, and more.

NSSN: A National Resource for Global Standards
www.nssn.org
Searchable database of national, regional and international standards.

The Online Ethics Center for Engineering and Science
http://onlineethics.org
Provides resources helpful in understanding and addressing ethically significant problems that arise in engineering.

Professional Publications
http://ppi2pass.com/catalog/servlet/MyPpi
Information on professional engineering, architecture, and ID exams.

Thomas Register of American Manufacturers
www.thomasregister.com
Directory of manufacturers that can be searched by company name, product or service, and brand name. Provides company description and contact information. Free registration required.

Virtual Library: Engineering
http://vlib.org/Engineering.html
A comprehensive link catalog for resources on engineering.

Vision Engineer
www.visionengineer.com/vision/main.shtml
Illustrated examples of engineering applications.

Organizations

Cornell University Engineering Library, Engineering Societies and Organizations
www.englib.cornell.edu/erg/soc.php
Directory of over 1,300 engineering organizations and includes their Web and e-mail addresses, description, list of publications, conferences, officers, and dues.

American Society for Engineering Education
www.asee.org
This organization is dedicated to promoting and improving engineering and technology education.

Standards: Agencies and Organizations
www.rmlibrary.com/db/stdagency.htm
Links to agencies and organizations that issue standards.

Subject Headings

- engineering
- engineering—encyclopedias
- engineering—handbooks, manuals, etc.

English Literature: A Research Guide

English literature is rich and varied and includes forms such as poetry, prose, drama, the novel, short stories, essays. As one of the oldest national literatures in the Western world, English literature dates back as far as A.D. 700. The materials listed below include information about authors from England, Scotland, and Wales and their works. This guide introduces researchers to some of the basic informational sources on the topic. The terms and phrases listed in the subject headings below can be used to search for more materials in the library's catalog and research databases. If you need further assistance, please ask a librarian.

Books

Annals of English Literature, 1475–1950 by Jyotish C. Ghosh, Elizabeth G. Withycombe, and Robert William Chapman. Oxford: Clarendon, 1961.
The principal publications of each year, together with an alphabetical index of authors with their works.

British Authors Before 1800 edited by Stanley Kunitz and Howard Haycraft. New York: H. W. Wilson, 1952.
Many of the entries in this alphabetically arranged resource are illustrated and nearly all end with a bibliography of works by and about the author.

British Authors of the Nineteenth Century edited by Stanley Kunitz and Howard Haycraft. New York: H. W. Wilson, 1936. ISBN: 0824200071.
A single volume biographical dictionary that provides brief accounts of the lives of the major and minor British authors of the 19th century.

British Writers by Ian Scott-Kilvert, 8 vols. New York: Charles Scribner's Sons, 1979–1984. ISBN: 0684157985.
An eight-volume set of literary biographies, originally published as individual studies in a British Council series called Writers and their Work.

The Cambridge Guide to English Literature by Ian Ousby. New York: Cambridge University Press, 1993. ISBN: 0521440866.
An illustrated guide to the literature of the English-speaking world covering more than 1,000 years.

A Literary History of England by Albert C. Baugh. New York: Appleton-Century-Crofts, 1967.
A comprehensive and scholarly history of the literature of England. Arranged by period and genre, with an extensive bibliographical supplement.

The New Cambridge Bibliography of English Literature edited by George Watson and Frederick W. Bateson. 5 vols. Cambridge: Cambridge University Press, 1969–1977. ISBN: 0521072557.
Covers English literature from 600 A.D. to modern times. Arranged chronologically and under periods by literary form. Vol. 1: 600–1660; vol. 2: 1660–1800; vol. 3: 1800–1900; vol. 4: 1900–1950; vol. 5: Index.

New Century Handbook of English Literature edited by Clarence L. Barnhart. New York: Appleton, 1967.
Over 14,000 entries with authors, titles, plots, characters, places, literary terms and allusions. Includes pronunciation keys and bibliographies.

Oxford Companion to English Literature edited by Margaret Drabble. New York: Oxford University Press, 2000. ISBN: 0198662440.
Includes brief articles on authors and literary works from all periods of English literature arranged alphabetically.

The Oxford Illustrated Literary Guide to Great Britain and Ireland by Dorothy Eagle, Hilary Carnell, and Meic Stephens. New York: Oxford University Press, 1993. ISBN: 019283133X.
Entries for places in Great Britain and Ireland with literary associations. Fictitious names of real places are entered as cross-references.

A Research Guide for Undergraduate Students: English and American Literature by Nancy L. Baker and Nancy Huling. New York: MLA, 2000. ISBN: 0873529782.
A concise guide for undergraduates for researching papers which provides a systematic way of locating materials and includes a selective bibliography.

Web Sites

The Camelot Project
www.lib.rochester.edu/camelot/cphome.stm
Sponsored by the University of Rochester, this database includes of Arthurian texts, images, bibliographies, and other related information.

English Literature on the Web
www.lang.nagoya-u.ac.jp/~matsuoka/EngLit.html
E-text archives, information on children's books and authors, medieval English literature, British authors and literature of various time periods.

Luminarium
www.luminarium.org/lumina.htm
This is an excellent resource for Middle English literature, Renaissance literature, and 17th-century literature.

Old English Literature
www.georgetown.edu/labyrinth/library/oe/oe.html
This site contains works such as *Beowulf* along with criticism about these works, as well as numerous prose, poems and reference works in Middle English.

Victoria Research Web
http://victorianresearch.org
Dedicated to the study of 19th-century British Literature this is a valuable resource for students, teachers and researchers studying the period.

Voice of the Shuttle—English Literature
http://vos.ucsb.edu
An excellent site on 18th-century literature and valuable for finding general resources, authors, works, projects, and criticism.

Subject Headings

- English literature—dictionaries
- English literature—encyclopedias
- English literature—16th century
- English literature—17th century
- English literature—18th century
- English literature—19th century
- English literature—20th century
- English literature—21st century

Environmental Studies: A Research Guide

Whether we live in the desert, rainforest, prairies, or near the ocean, our environment is consistently at risk due to the actions of man and nature. The resources below will help researcher get started in learning more about the concepts and issues facing our environment today. This guide introduces researchers to some of the basic informational sources on the topic. The terms and phrases listed in the subject headings below can be used to search for more materials in the library's catalog and research databases. If you need further assistance, please ask a librarian.

Reference Books

The Dictionary of Ecology and Environmental Science by Henry W. Art. New York: Henry Holt, 1993. ISBN: 0805020799.
Over 8,000 entries with definitions including concepts from environmental biology, chemistry, geology, and physics.

Encyclopedia of Environmental Science edited by John F. Mongillo and Linda Zierdt-Warshaw. Phoenix: Oryx Press, 2000. ISBN: 1573561479.
Covers basic terminology and key topics in the field of environmental science.

Environmental Encyclopedia, 3rd ed. edited by Marci Bortman, et al. 2 vols. Detroit: Gale Group, 2003. ISBN: 0787654868.
Consists of nearly 1,300 signed articles and term definitions provides in-depth, worldwide coverage of environmental issues.

Environmental Literacy: Everything You Need to Know About Saving Our Planet by H. Steve Dashefsky. New York: Random House, 1993. ISBN: 0679412808.
Alphabetical listing of key words, expressions, and concepts intended to familiarize the nonspecialist with current issues, concepts, and environmental themes.

The Facts On File Dictionary of Environmental Science by Bruce C. Wyman and L. Harold Stevenson. New York: Facts On File, 2001. ISBN: 0816042330.
Covers environmental issues such as contamination of air and water, natural resources conservation, and workplace health and safety in over 3,000 entries.

The Green Encyclopedia by Irene M. Franck and David M. Brownstone. New York: Prentice Hall General Reference, 1992. ISBN: 0133656853.
Single volume that covers such varied environmental themes from environmental disasters to dangerous pesticides to threatened species. Includes 50 line drawings.

Macmillan Encyclopedia of the Environment edited by Stephen R. Kellert, et al. 6 vols. New York: Macmillan Library Reference, 1997. ISBN: 002897381X.
Provides basic information about such topics as minerals, energy resources, pollution, soils and erosion, wildlife and extinction, agriculture, the ocean, wilderness, hazardous wastes, population, environmental laws, ecology, and evolution.

McGraw-Hill Encyclopedia of Environmental Science. New York, McGraw-Hill, 1974. ISBN: 0070452601.
Over 300 signed and illustrated articles for the nonspecialist about environmental science. Includes cross-references, bibliographies, index, maps, photographs, drawings, and charts.

Web Sites

Librarians Index to the Internet, Environment
http://lii.org/search/file/environ
Internet search directory that provides links to reliable Web sites on many varied topics from endangered species and noise pollution to mercury contamination and solar energy.

The National Environmental Directory
www.environmentaldirectory.net
Directory of more than 13,000 organizations in the United States concerned with environmental issues and environmental education.

National Library Environment
www.ncseonline.org/NLE/
Offers access to over 1,000 Congressional research service reports on the environment and related topics.

Terms of Environment, U.S. Environment Protection Agency (EPA)
www.epa.gov/OCEPAterms/intro.htm
Defines in nontechnical language the more commonly used environmental terms appearing in EPA publications.

Organizations

Center for Environmental Information (CEI)
www.rochesterenvironment.org
Private, nonprofit, educational organization that provides information and communication services, publications, and educational programs.

Conservation International (CI)
www.conservation.org
Preserves and promotes awareness about the world's most endangered biodiversity through scientific programs, local awareness campaigns, and economic initiatives.

Environmental Protection Agency
www.epa.gov
Provides a wide variety of information on the air, conservation, ecosystems, health and safety, waste and recycling, and water and environmental industry news. Includes areas specifically designed for children and high school students.

Kids for a Clean Environment (Kids FACE)
www.kidsface.org
Organization comprised of children, parents, teachers, and others working to improve the environment. Focus is on children organizing and implementing ideas and programs on their own, supported and assisted by parents and teachers.

Subject Headings

- environmental engineering
- environmental health
- environmental impact analysis
- environmental monitoring
- environmental protection
- environmental sciences
- environmentally induced diseases
- health risk assessment
- human ecology
- nature—effect of human beings on
- pollution—environmental aspects

Etiquette: A Research Guide

Some of the questions that this pathfinder will answer will include: Should I bring my hosts a gift of liquor when I'm a guest for the weekend? What do I say when someone greets me? How quickly do I need to write that thank-you note to Aunt Dottie? What was etiquette for "going Dutch" on a date in the 1950s? And, of course...which fork do I use? Materials listed in this guide focus primarily on home etiquette—rules for behavior at a table or when you are a guest—rather than business etiquette. Additionally, the materials listed below focus on the traditional etiquette of Europe and the United States. The terms and phrases listed in the subject headings below can be used to search for more materials in the library's catalog and research databases. If you need further assistance, please ask a librarian.

Etiquette Classics

Complete Book of Etiquette by Nancy Tuckerman, Nancy Dunnan, and Amy Vanderbilt. New York: Doubleday, 1995. ISBN: 0385413424.
First published in 1952 and last revised in 1978, is as much a guide to contemporary living as it is an etiquette book. Updated version tells reader where old expectations still apply and where customs have changed.

Emily Post's Etiquette, 16th ed. by Emily Post and Peggy Post. New York: HarperCollins, 1997. ISBN: 0062700782.
Thoroughly revised and updated, *Emily Post's Etiquette* now includes etiquette surrounding cultural diversity, birth and death ceremonies of many religions, and etiquette for international travelers. Still includes all of the etiquette rules that have made this a hallmark resource.

Miss Manners' Guide to Excruciatingly Correct Behavior by Judith Martin. New York: Atheneum, 1982. ISBN: 0689112475.
Covers topics like courtesy, proper attire, and the etiquette of weddings. Martin knows right from wrong and sensible from rude. Delivered with comical insight.

Etiquette for Adults

These guides cover the basics of etiquette for adults (although some of them include sections on manners for kids).

Advice Lady
http://advicelady.com/answerpage.asp?category=Manners.
Advice on manners from a nonprofessional. Read others' questions and the Advice Lady's answers, or ask her your own.

About: Etiquette for Personal Entertaining
http://entertaining.about.com/od/etiquetteforentertaining/
A portal site leading to a variety of other etiquette resources online.

Etiquette for Kids and Teens

These books are intended for children and teenagers. They generally cover both the "how-tos" and the "whys" of table manners.

Be the Best You Can Be: A Guide to Etiquette and Self-Improvement for Kids and Teens by Robin Thompson. Pekin, Illinois: Robin Thompson Charm School, 1999. ISBN: 0967531802.
A wealth of information just for this age group on poise and confidence, positive attitude, shyness, social skills, and much more.

Let's Talk About Good Manners by Diana Shaughnessy. New York: PowerKids Press, 2003. ISBN: 082395045X.
A simple discussion of what good manners are, why good manners are important, and how manners are different in different cultures.

Historical Etiquette

These are both etiquette books from the past and books about the etiquette of the past.

Co-Ediquette: Poise and Popularity for Every Girl by Elizabeth Eldridge. New York: E. P. Dutton, 1936.

George Washington's Rules of Civility: 110 Maxims Helped Shape and Guide America's First President.
www.npr.org/features/feature.php?wfId=1248919
Rules of civility and decent behavior in company and conversation. Reprinted on National Public Radio's Web site.

Miss Abigail's Time Warp Advice
www.missabigail.com
A potpourri of advice from the past century or so, presented in Q&A style.

Post, Emily. 1922. Etiquette in Society, in Business, in Politics and at Home
www.bartleby.com/95/index.html
The original text of Post's *Etiquette in Society, in Business, in Politics and at Home.*

The Rudiments of Genteel Behavior by Francois Nivelon. Seattle: University of Washington Press, reprint ed., 2003, 1737. ISBN: 1903470102.
This 18th-century book of manners is a must-read for historians, actors and dancers.

The Seventeen Guide to Your Widening World by Enid Annenberg Haupt. New York: Macmillan, 1965.

Subject Headings

If you would like to seek out books on etiquette at your local library, you can try looking under some of the following Library of Congress Subject Headings (these can usually be searched through in the online card catalog at your library). The heading Etiquette can be subdivided geographically, so if you are looking for etiquette literature on other cultures, you can look under Etiquette—Japan or other geographic region.

- etiquette
- etiquette—bibliography
- etiquette—United States
- Europe—social life and customs
- United States—social life and customs

Exploration and Discovery: A Research Guide

Learn more about voyages to the bottom of the sea, adventures in space, and travels across the globe, both historic and contemporary. This guide contains books, Web sites, and other materials to introduce you to historic explorers you may already know like Christopher Columbus or some explorers of today. It might even help you figure out if you have what it takes to be an adventurer yourself. This guide introduces researchers to some of the basic informational sources on the topic. The terms and phrases listed in the subject headings below can be used to search for more materials in the library's catalog and research databases. If you need further assistance, please ask a librarian.

Books

Against all Opposition: Black Explorers in America by James Haskins. New York: Walker, 1992. ISBN: 0802781381.
Surveys the lives and adventures of black explorers who helped discover new worlds.

American Heroes of Exploration and Flight by Anne E. Schraff. Springfield, NJ: Enslow, 1996. ISBN: 0894906194.
Biographies of American adventurers including Robert Peary, Amelia Earhart, Neil Armstrong, and Christa McAuliffe.

Career Ideas for Kids Who Like Adventure by Diane L. Reeves, Nancy Heubeck, and Nancy Bond. New York: Facts On File, 2001. ISBN: 0816043213.
A guide through a multitude of career possibilities based on specific interests and skills. Links talents to a wide variety of actual professions.

DK Illustrated Book of Great Adventurers: Real-life Tales of Danger and Daring by Richard Platt. New York: Dorling Kindersley, 1999. ISBN: 0789444615.
Presents activities and accomplishments of great adventurers, including explorer Leif Eriksson, the glamorous spy Mata Hari, and astronaut Neil Armstrong.

Explorers of North America by Brendan January. New York: Children's Press, 2000. ISBN: 0516216295.
Describes activities of explorers in North America from 1000 to 1804 including Leif Ericsson, Christopher Columbus, Amerigo Vespucci, and Lewis and Clark.

Exploring the New World by Rebecca Stefoff. Tarrytown, NY: Benchmark Books, 2001. ISBN: 0761410562.
Traces exploration of America from the arrival of Asian hunters in Alaska more than 15,000 years ago to the English and French settlers in the early 1600s.

Land Ho!: Fifty Glorious Years in the Age of Exploration with 12 Important Explorers by Nancy W. Parker. New York: HarperCollins, 2001. ISBN: 0060277599.
Shows how the voyages of Columbus, Cabot, Ponce De Leon, and other explorers to the Americas were the result of mistakes, accidents, and misses.

Living Dangerously by Doreen Rappaport. New York: HarperCollins, 1991. ISBN: 0060251085.
Features American women who risked their lives for adventure.

The Seven Seas: Exploring the World Ocean by Linda Vieira and Higgins Bond. New York: Walker, 2003. ISBN: 0802788335.
Explores the diversity of the oceans, seas, and gulfs of the planet Earth.

The Sky's the Limit: Stories of Discovery by Women and Girls by Catherine Thimmesh and Melissa Sweet. Boston: Houghton Mifflin, 2002. ISBN: 0618076980.
Brief accounts of the work of a variety of women scientists in such fields as astronomy, biology, anthropology, and medicine.

Space Exploration by Dana M. Rau. Minneapolis: Compass Point Books, 2003. ISBN: 0756504392.
Looks at the universe as it describes telescopes as important tools, manned and unmanned missions, and discusses the future of space exploration.

A World of Wonders: Geographic Travels in Verse and Rhyme by J. Patrick Lewis and Alison Jay. New York: Dial Books for Young Readers, 2002. ISBN: 0803725795.
Through illustrated poems kids can travel the globe to learn amazing facts about spaces and places in the wonderful world of geography.

Audio Visual

History of Exploration. Wynnewood, PA: Schlessinger Media, 2000. ISBN: 1572253363. (VHS 23 min.)
Shows the motivations of early explorers whose bravery was the foundation for the successes during the Golden Age of Discovery. Grades 5–8.

Space Exploration. Wynnewood, PA: Schlessinger Media, 1999. ISBN: 1572252413. (VHS 26 min.)
Covers major challenges and benefits of exploration, including rocket science, a look back at the "Space Race," and a history of manned space travel. Grades 5–8.

Web Sites

Discovery Channel
http://dsc.discovery.com/convergence/quest/quest.html
Quest partners with scientists and explorers at the vanguard of their fields and follows them to see their far-reaching impact on our understanding of the world.

Library of Congress: Discovery and Exploration
http://memory.loc.gov/ammem/gmdhtml/dsxphome.html
Documents discovery and exploration using manuscripts and maps. Maps from the late 15th century to the 19th century by Lewis and Clark and others.

National Aeronautics and Space Administration (NASA)
www.nasa.gov/missions/solarsystem/explore_main.html
Learn about NASA, the missions, space exploration history, research aircraft, and much more.

NOVA/PBS Online Adventures
www.pbs.org/wgbh/nova/adventures/
Series of real-time expeditions follows scientists and explorers into the field, reporting on science as it happens and allowing audience participation via e-mail.

Voyage of Exploration
http://library.thinkquest.org/C001692/english/index.php3?subject=home
Exploration from 1500 B.C. to present. Includes information on explorers; expeditions; maps; quizzes, and time line. Also "Teachers' World" with lessons.

Subject Headings

- adventure and adventurers
- adventure and adventurers—juvenile literature
- discoveries in geography
- explorers
- explorers—juvenile literature

Financial Aid and Scholarships: A Research Guide

As college costs continue to rise, the search for financial aid becomes more competitive. Students familiar with the sources of aid and the process by which to apply for it will find themselves at an advantage when it comes time to pay the bursar. Thousands of sources exist but only those students who fit the criteria are eligible to apply. This guide introduces researchers to some of the basic informational sources on the topic. The terms and phrases listed in the subject headings below can be used to search for more materials in the library's catalog and research databases. If you need further assistance, please ask a librarian.

Directories

The College Board Scholarship Handbook 2004, 7th ed. College Entrance Examination Board. New York: College Board, 2003. ISBN: 0874476984.
Guide to over 2,000 scholarships, internships, and loan programs for undergraduates.

Financial Aid for… Lists. El Dorado Hills, CA.: Reference Service Press, 1997– . Biennial.
A list of scholarships, fellowships, loans, grants, awards, and internships open primarily or exclusively to African-Americans, Asian-Americans, Hispanic-Americans, and Native Americans.

Financial Aid for African Americans, ISSN: 1099906X.

Financial Aid for Asian Americans, ISSN: 10999124.

Financial Aid for Hispanic Americans, ISSN: 10999078.

Financial Aid for Native Americans, ISSN: 10999116.

Scholarships, Fellowships, and Loans edited by S. Norman Feingold and Marie Feingold. Boston: Bellman, 2004– . Annual. ISSN: 0787688207.
Provides more than 3,700 sources of education-related financial aid and awards at all levels of study.

Web Sites

College Board Online
www.collegeboard.com
Site is designed for teachers, educators, and parents. Offers information about colleges, admission tests, and scholarship information.

Collegenet
www.collegenet.com
Search for colleges, scholarships, and other college resources.

FastWeb
www.fastweb.com
College and scholarship search.

Federal Student Aid
http://studentaid.ed.gov
Provides resources to take you through steps to college success including: preparing for college, choosing the right school, applying to schools, finding funding to attend school, and repaying college debt.

Finaid
www.finaid.org
Established in the fall of 1994 as a public service. This award-winning site has grown into the most comprehensive annotated collection of information about student financial aid on the Web.

Free Application for Federal Student Aid, U.S. Department of Education
www.FAFSA.ed.gov
To apply for federal student financial aid, and to apply for many state student aid programs, students must complete a "Free Application for Federal Student Aid" (FAFSA).

IPEDS College Opportunities Online
http://nces.ed.gov/IPEDS/cool/Search.asp
Includes many ways to search nearly 7,000 colleges and universities in the United States by size, specialization, careers or trades.

Scholarship Scams, Federal Trade Commission
www.ftc.gov/bcp/conline/edcams/scholarship/index.html
Learn tips on how to avoid scholarship scams.

Student Gateway to the United States Government
www.students.gov
Offers a variety of information for students including how to find the right college, apply for federal student financial aid, learn about military educational benefits, find a job or internship, and register for selective service.

The Student Guide
http://studentaid.ed.gov/students/publications/student_guide/2004_2005/english/index.htm
This guide explains student financial aid programs the U.S. Department of Education's Federal Student Aid (FSA) office administers.

Student Resources, Office of Postsecondary Education, U.S. Department of Education
www.ed.gov/about/offices/list/ope/students.html
Lists financial assistance and scholarship programs for undergraduates and graduates.

"Winning the Financial Aid Game" by Sarah Breckenridge. *Smart Money*, October 30, 2002.
www.smartmoney.com/mag/index.cfm?story=nov02-college
Article details the college entrance and scholarship process with tips on how to find more money.

Subject Headings

Add the terms "directories" or "periodicals" to the subject headings below to narrow your search results.

- African American college students—scholarships, fellowships, etc.
- Asian American college students—scholarships, fellowships, etc.
- Federal aid to higher education—United States
- fellowships and scholarships—United States
- Hispanic American college students—scholarships, fellowships, etc.
- Indian college students—scholarships, fellowships, etc.
- internship programs—United States
- minorities—scholarships, fellowships, etc.—United States
- scholarships—United States
- student aid—United States
- training support—United States

Finding Forms: A Research Guide

Chances are that you will need to find or create a form for personal or business reasons at some point, whether it is to transfer title of a vehicle, prepare your income taxes, or rent an apartment. The resources below point to free resources available in your library or on the Internet. Additionally, you may find forms included in books on the very topic of the form. For example, sample forms may be included in books about real estate, divorce, income taxes, or just about any topic requiring forms. This guide introduces researchers to some of the basic informational sources on the topic. The terms and phrases listed in the subject headings below can be used to search for more materials in the library's catalog and research databases. If you need further assistance, please ask a librarian.

Books

101 Law Forms for Personal Use, 3rd ed. by Ralph E. Warner and Robin Leonard. Berkeley, CA: Nolo Press, 2002. ISBN: 0873378466.
Includes sample forms including delegating authority to care for children, pets and property, basic estate planning, renting residential real estate, borrowing or lending money, and much more.

Business Forms on File, 2004 ed. 2 vols. New York: Forms On File, 2004. ISBN: 0816055890.
Organized by subject, this volume contains more than 140 of the forms most frequently used in business in areas such as accounting and finance, government aid programs, personnel, government procurement programs, and real estate.

The Complete Book of Small Business Legal Forms, 3rd ed. by Dan Sitarz. Carbondale, IL: Nova, 2001. ISBN: 0935755845.
Contains all of the forms and instructions that a start-up entrepreneur will need to ensure business success. Forms also included on CD-ROM.

The Most Valuable Personal Legal Forms You'll Ever Need by Mark Warda and James C. Ray. Naperville, IL: Sphinx, 2001. ISBN: 1572481307.
Forms include deeds, living wills, promissory notes, wills, power of attorney, and living trusts.

Personal Forms on File, 2004 ed. 2 vols. New York: Forms On File, 2004. ISBN: 0816055882.
Includes personal forms in areas such as financial aid, education, financial and personal property, health and social services benefits, legal military and veterans' affairs, and real estate and rental.

Web Sites

Court Forms
http://forms.lp.findlaw.com
The federal forms collection contains court forms for federal appellate, district and bankruptcy courts, and court forms for state courts.

Federal Forms
www.irs.ustreas.gov
Forms and publications are available for download through the Internal Revenue Service.

Federation of Tax Administrators
www.taxadmin.org/fta/link/forms.html
Includes links to state income tax forms.

Fedforms
www.fedforms.gov
FedForms.gov provides "one-stop shopping" for the Federal forms most used by the public.

Forms from the Feds
http://exlibris.memphis.edu/resource/unclesam/forms.html
Provides links to the most frequently requested federal government forms. List comprises the most frequently requested forms.

The 'Lectric Law Library Forms Room
www.lectlaw.com/form.html
Provides many sample forms including business and general topics dealing with subjects like real estate, corporations, employment, contracts, trusts and wills.

State Corporation and Business Forms
www.findlaw.com/11stategov/indexcorp.html
Includes links to business forms.

Subject Headings

- business—forms
- contracts—United States—forms
- contracts—United States—popular works
- estate planning—United States—forms
- filing systems
- forms (law)—United States—popular works
- probate law and practice—United States—forms
- records—forms

Finding People: A Research Guide

The resources listed below will get you started on your way to finding old friends, lost family, high school sweethearts and military buddies. Hundreds of free information sources exist to unite you—it's just knowing that they exist and how to access them. Many of the guides and Web sites below will identify online sources or more traditional methods of using the phone and postal mail to find just about anyone. The terms and phrases listed in the subject headings below can be used to search for more materials in the library's catalog and research databases. If you need further assistance, please ask a librarian.

Books

Find Anyone Fast, 3rd ed. by Richard S. Johnson and Debra Johnson Knox. Spartanburg, SC: MIE, 2001. ISBN: 1877639850.
Included are directories of telephone number, voter registration records, death records, and listings for lawyers, doctors, pilots, members of the military, and more. Emphasis on online sources of information.

How to Locate Anyone Anywhere Without Leaving Home, rev. ed. by Ted L. Gunderson. New York: Plume, 1996. ISBN: 0452277426.
Methodical guide to locate missing persons by mail or telephone using hundreds of sources.

How to Locate Anyone Who Is or Has Been in the Military: Armed Forces Locator Guide, 8th ed. by Richard S. Johnson and Debra Johnson Knox. Spartanburg, SC: MIE, 1999. ISBN: 1877639508.
Provides sources of information to assist in searching for former military personnel.

Public Records Online: The National Guide to Private & Government Online Sources of Public Records, 4th ed. edited by Michael L. Sankey and Peter J. Weber. Tempe, AZ: Facts on Demand Press, 2003. ISBN: 1889150371.
Lists thousands of online resources for private and government public information such as address information, bankruptcy, genealogy, real estate, vital records, trade names, tax liens, and fictitious names through county, state and federal government agencies.

Public Records

National Association of Counties
www.naco.org, About Counties, Find a County
Links to county sites. Search for assessor pages, records clerk, or other departments that may gather information on people.

State and Local Government on the Net
www.statelocalgov.net
Provides links to local municipalities on the Web. Click through to the town clerk's office or the assessor's page to learn more about accessing property information or vital records.

Where to Write for Vital Records
www.cdc.gov/nchs/howto/w2w/w2welcom.htm
To use this valuable tool, you must first determine the State or area where the birth, death, marriage, or divorce occurred, then click on that State or area.

Telephone Directories: Yellow Pages, White Pages and Reverse Lookups

Many directories include search by person's or company's name, yellow page heading, or reverse telephone number. These directories may provide postal addresses, telephone numbers and e-mail addresses for people.

411locate: *www.411locate.com*
AnyWho Online Directory: *www.anywho.com*
Infobel Worldwide Directories: *www.infobel.com/teldir/*

InfoSpace: *www.infospace.com*
International White Page Phone Directory: *www.whitepages.com/intl_sites.pl*
The Ultimates: *www.theultimates.com*
Yahoo! People Search: *http://people.yahoo.com*

Special Directories

Many times you can locate someone because they are in a certified profession like accountancy, dentistry, or cosmetology. Many times they are a member of an association that publishes its membership directory on the Web. The resources listed below can help get you started. For more information search the following keywords in the library catalog or your favorite search engine: (profession) and directories (i.e., dentist and directories).

Associations on the Net
www.ipl.org/div/aon/
Search by keyword to identify an association. Many times member directories are published on the Web.

People Spot, Special Directories
www.peoplespot.com/directories/
Lists links to Web pages that provide directories of business executives, dentist, physician, real estate agents, veterinarian, and more.

Web Sites

National Adoption Directory: State Reunion Directories
http://naic.acf.hhs.gov/database/nadd/naddsearch.cfm
Search by state to find out if yours has a state reunion directory. Allows parties of adoption to enter their names on a central registry, where the two parties may be matched and placed in contact with one another.

PeopleSpot.com
www.peoplespot.com
Large portal site with links to hundreds of relevant sites.

Subject Headings

- missing persons—investigation—United States
- missing persons—investigation—United States—handbooks, manuals, etc
- online information services—United States—directories
- public records—United States—directories

Gardening: A Research Guide

Finding information on gardening, the practical operations in the cultivation of fruits, vegetables, flowers and ornamental plants, is easier than ever. The resources listed below are works that are not just meant to be consulted for reference, but also to be read and enjoyed by the casual reader with illustrations being an important part of these sources. The topics range from the history of gardening to specific information on plants and their cultivation to garden and landscape planning. This guide introduces researchers to some of the basic informational sources on the topic. The terms and phrases listed in the subject headings below can be used to search for more materials in the library's catalog and research databases. If you need further assistance, please ask a librarian.

Books

American Horticultural Society Encyclopedia of Gardening by Christopher Brickell. New York: Dorling Kindersley, 2003. ISBN: 0789496534.
Newest revision of this long-standing, comprehensive gardening resource.

American Horticultural Society Encyclopedia of Plants and Flowers by Christopher Brickell and Trevor J. Cole. New York: Dorling Kindersley, 2002. ISBN: 0789489937.
Includes great illustrations with sections on plant selection, a plant catalog, and dictionary.

The American Horticultural Society Gardening Manual. New York: Dorling Kindersley, 2000. ISBN: 0789459523.
Covers every aspect of gardening from preliminary design to daily maintenance.

Annuals by James Crockett. New York: Time-Life Books, 1971. LCCN: 78140420.
Part of the Time-Life Encyclopedia of Gardening series featuring watercolor illustrations of annuals.

Better Homes and Gardens Complete Guide to Gardening by Marjorie P. Groves. Des Moines: Meredith, 1979. ISBN: 0696000415.
Technical and detailed overview of all aspects of gardening. The 1994 edition is also on CD-ROM.

The National Arboretum Book of Outstanding Garden Plants: The Authoritative Guide to Selecting and Growing the Most Beautiful, Durable, and Care-Free Garden Plants in North America by Jacqueline Hériteau and Henry M. Cathey. New York: Simon & Schuster, 1990. ISBN: 0671669575.
This book focuses on selection of the right plant for a specific location.

New Encyclopedia of Herbs & Their Uses by Deni Bown. New York: Dorling Kindersley, 2001. ISBN: 078948031X.
Definitive A-Z guide to growing herbs around the globe.

The New York Botanical Garden Illustrated Encyclopedia of Horticulture by Thomas H. Everett. New York: Garland, 1980–1982. ISBN: 0824063080 (vol. 1).
Extensive general and specific information on horticulture useful to any gardener.

The Oxford Companion to Gardens by Patrick Goode and Michael Lancaster. New York: Oxford University Press, 1986. ISBN: 0198661231.
Covers the history and design of gardens from the earliest known examples to the present day.

Rodale's All-New Encyclopedia of Organic Gardening: The Indispensable Resource for Every Gardener by Fern Marshall Bradley and Barbara W. Ellis. Emmaus, PA: Rodale Press, 1992. ISBN: 0878579990.
A basic reference for organic and other gardening with over 400 entries covering composting, xeriscaping, permaculture, and much more.

Web Sites

Burpee Seeds and Plants
www.burpee.com
Seller of specialty seeds and gardening accessories.

The Garden Gate: Into Gardening and Beyond
http://garden-gate.prairienet.org
From Forbes Best of the Web, "This folksy link site earns high marks for its breadth, organization, and helpful annotations."

Garden Web: The Internet's Garden Community
www.gardenweb.com
Site provides links to a variety of gardening resources including forums, "HortiPlex Plant Database," calendar of garden events, gardening organizations, glossary of terms, and much more.

GardenGuides
www.gardenguides.com
Features an article index on a variety of gardening topics and links to many other gardening sites and guides. Registration for forums is required but free.

Gardening Launch Pad
www.gardeninglaunchpad.com
Almost 5,000 links to gardening sites with 95% content links not commercial.

HGTV Home & Garden Television
www.hgtv.com/hgtv/gardening/
This well-known television network's site is a comprehensive resource for gardening information.

United States Department of Agriculture: Home Gardening
www.usda.gov/news/garden.htm
Links to everything from backyard conservation to the USDA's Vegetable Laboratory.

Subject Headings

- agriculture
- gardening
- herb gardening
- herbs
- horticulture
- landscape gardening
- medicine, herbal
- plants, cultivated
- plants, medicinal
- plants, ornamental

Gay and Lesbian Studies: A Research Guide

Materials listed below will assist the beginning researcher in locating information that helps provide background information on many aspects of social, political, cultural, psychological, and historical aspects of issues affecting gay, lesbian, bisexual and transgender people. The subject headings listing at the end of this guide can be searched in the library catalog in order to locate more information on the topic.

Reference Books

Completely Queer: The Gay and Lesbian Encyclopedia by Steve Hogan and Lee Hudson. New York: Henry Holt, 1998. ISBN: 0805036296.
Reference compendium including facts, reading lists, and useful tables from famous pseudonyms to gay detective novels. Covers both lesbian and gay male points of view.

Encyclopedia of Homosexuality by Wayne R. Dynes, Warren Johansson, William A. Percy, and Stephen Donaldson. New York: Garland, 1990. ISBN: 0824065441.
Includes topical entries on the historical, medical, psychological, and sociological aspects of homosexuality. Biographies exclude living people. Bibliography included.

Encyclopedia of Lesbian, Gay, Bisexual and Transgendered History in America edited by Marc Stein. New York: Charles Scribner's Sons, 2003. ISBN: 0684312611.
Provides a comprehensive survey of lesbian and gay history and culture in the United States. Includes 545 articles ranging from short biographical entries to longer essays, as well as a guide to archival sources, a chronology/time line, a historical overview essay and a comprehensive index.

Encyclopedia of Lesbian and Gay Histories and Cultures. 2 vols.
Lesbian Histories and Cultures. Vol. 1 edited by Bonnie Zimmerman. ISBN: 0815319207; Gay Histories and Cultures. Vol. 2 edited by George E. Haggerty, et al. New York: Garland, 2000. ISBN: 0815318804.
Each volume consists of brief entries arranged alphabetically. Brief bibliographies, subject guides, cross-references, and indexes help locate more information.

Historical Dictionary of the Gay Liberation Movement: Gay Men and the Quest for Social Justice by Ronald J. Hunt. Lanham, MD: Scarecrow Press, 1999. ISBN: 0810835878.
Historical overview of the gay liberation movement.

St. James Press Gay & Lesbian Almanac edited by Neil Schlager. Detroit: St. James Press, 1998. ISBN: 1558623582.
Details gay and lesbian history, culture, and experience in 20th-century America.

Journal Titles

The Advocate. Los Angeles: Liberation, 1970– . Biweekly. ISSN: 00018996.
National gay and lesbian news magazine.

Journal of Homosexuality. New York: Haworth Press, 1974– . Quarterly. ISSN: 00918369.
Devoted to scholarly research on homosexuality, including sexual practices and gender roles and their cultural, historical, interpersonal, and modern social contexts.

Web Sites

glbtq: gay, lesbian, bisexual, transgender, and queer culture
www.glbtq.com
This Web-based encyclopedia provides scholarly entries within the arts, literature, and social sciences. Researchers may browse or search to retrieve relevant information. Bibliographies and extensive citations are included with entries.

Out of the Past
www.pbs.org/outofthepast/
Documentary includes facets of our history that have been left out of the textbooks. Site continues the exploration begun by film of same title.

Stateline.org: 50-State Rundown on Gay Marriage Laws
www.stateline.org
The introductory paragraphs of this article provide an overview of recent changes in gay marriage laws. A state-by-state listing of the current law and legislative background follows.

Organizations

Gay and Lesbian Alliance Against Defamation
www.glaad.org
Promotes fair, accurate, and inclusive representation of people and events in the media as a means of eliminating homophobia and discrimination based on gender identity and sexual orientation.

Human Rights Campaign
www.hrcusa.org
Lobbies Congress, provides campaign support, and educates the public about issues including workplace, family, discrimination and health issues for gay, lesbian, bisexual and transgender people.

Lambda Legal Defense and Education Fund
www.lambdalegal.org
Supports civil rights of lesbians, gay men, bisexuals, transgender people, and people with HIV or AIDS through litigation, education, and public policy.

The National Gay and Lesbian Task Force
www.ngltf.org
Supports civil rights of gay, lesbian, bisexual, and transgender people through political movement.

PFLAG (Parents and Friends of Lesbians and Gays)
www.pflag.org
PFLAG is a national nonprofit organization with over 200,000 members.

Subject Headings

- gay liberation movement—United States—periodicals
- gay men—United States
- homosexuality—dictionaries
- homosexuality—encyclopedias
- homosexuality—United States—periodicals
- lesbianism—encyclopedias
- lesbians—encyclopedias
- lesbians—United States
- transgender (transsexualism)

Genealogy: A Research Guide

Researching one's family history can be a lifelong journey. Uncovering records and documents that help to piece together the past lives of ancestors is a labor of love for many. If you are new to genealogy research this guide will provide you with a great place to start. The resources listed below present the most standard resources in the field. Many of them will point you to even more resources. If you need assistance in locating books that the library does not own the librarian can help you find it at another library. The terms and phrases listed in the subject headings below can be used to search for more materials in the library's catalog and research databases. If you need further assistance, please ask a librarian.

Books

Ancestral Trails: The Complete Guide to British Genealogy and Family History by Mark D. Herber. Baltimore: Genealogical, 1998. ISBN: 0806315415.
Guide to tracing British records outlines basic steps such as drawing family trees, using census records, and searching for birth, marriage, and death certificates.

Family Pride: The Complete Guide to Tracing African-American Genealogy by Donna Beasley. New York: Macmillan General Reference, 1997. ISBN: 0028608429.
Guide to researching African-American family history and genealogy and offers step-by-step instructions on how to conduct an effective search.

The Genealogist's Companion & Sourcebook, 2nd ed. by Emily A. Croom. Cincinnati: Betterway, 2003. ISBN: 1558706518.
A beyond-the-basics guide to tracing family history. Explains how to research, locate, and use church and funeral home records; government records at federal, state, and local levels, including the U.S. Serial Set and the Territorial Papers of the United States; court records; newspapers; and maps.

Genealogy Online: Researching Your Roots, 7th ed. by Elizabeth P. Crowe. New York: McGraw-Hill, 2003. ISBN: 0072229780.
Directory of genealogy Web sites, newsgroups, mailing lists, and commercial services.

The Genealogy Sourcebook by Sharon D. Carmack. Los Angeles: Lowell House, 1998. ISBN: 1565657942.
Discusses various documentation methods, where and how to find records, and organization methods.

Printed Sources: A Guide to Published Genealogical Records by Kory L. Meyerink. Salt Lake City, UT: Ancestry, 1998. ISBN: 0916489701.
Focuses on secondary sources, including encyclopedias, gazetteers, indexes, abstracts, histories, biographies, military sources, and periodicals. Appendixes include CD-ROMs for family historians, major genealogical libraries in the United States, and a list of genealogical publishers and booksellers.

The Researcher's Guide to American Genealogy, 3rd. ed. by Val D. Greenwood. Baltimore: Genealogical, 2000. ISBN: 0806316217.
First part provides the foundation of genealogical research; the second part covers records and their use.

Unpuzzling Your Past: A Best-Selling Basic Guide to Genealogy, 4th ed. by Emily A. Croom. Cincinnati: Betterway, 2003. ISBN: 1558705562.
Includes chapters on how to get started, the meaning of names, the difference between a family history of dates and a family history of stories, how to gather sources, who to interview, and how to fit it all together.

Web Sites

Ancestry.com Genealogy and Family History Records
www.ancestry.com
Provides many genealogy databases, worldwide resources and the Eastman Online Genealogy Newsletter. Some information available by subscription.

Cyndi's List of Genealogy Sites on the Internet
www.cyndislist.com
Comprehensive list of genealogy Web sites with thousands of links according to categories for easy reference. Includes such topics as "How to Begin," "Tools for Research," and "Publishing Your Family History."

The Family TreeMaker Online
www.familytreemaker.com
Offers professional research services, message boards for genealogists, and searchable databases. Requires a subscription fee.

FamilySearch Internet Genealogy Service
www.familysearch.org
The Church of Jesus Christ of Latter-Day Saints provides this site, which lets you search their "Ancestral File," "International Genealogical Index," "Pedigree Resource File," and Web sites (by last name only).

The Genealogy Home Page
www.genhomepage.com
Organized into topics such as "Genealogy Help and Guides," "Libraries," "Maps," "Newsgroups and Genealogy Societies." The Genealogy Home Page also has a list of new genealogy links that is updated daily.

RootsWeb Genealogical Data Cooperative
www.rootsweb.com
Many databases, services, and downloadable files are available only through subscription.

The U.S. Genealogy Web Project
www.usgenweb.com
Volunteers working to create a center for genealogical research for every county in the United States. All counties have an area for you to post queries and links to the state page and archives.

Subject Headings

- African Americans—genealogy—handbooks, manuals, etc.
- genealogy
- Great Britain—genealogy—handbooks, manuals, etc.
- United States—genealogy—bibliography
- United States—genealogy—computer network resources
- United States—genealogy—handbooks, manuals, etc.

Geography: A Research Guide

Geography describes and analyzes the spatial variations in physical, biological, and human phenomena that occur on the surface. It is the scientific study of the Earth's surface. The resources listed below provide students and researchers with the basic introductory material in the field. If you need assistance in locating books that the library does not own the librarian can help you find it at another library. The terms and phrases listed in the subject headings below can be used to search for more materials in the library's catalog and research databases. If you need further assistance, please ask a librarian.

Books

Biographical Dictionary of Geography by Robert P. Larkin and Gary L. Peters. Westport, CT: Greenwood Press, 1993. ISBN: 0313276226.
Examines a cross section of geographers, from ancient to modern times. Entries contain selected bibliographies of works by and about the person.

The Columbia Gazetteer of the World by Saul B. Cohen. New York: Columbia University Press, 1998. ISBN: 0231110405.
Over 160,000 entries on places in the world including economic, political, historical, cultural descriptions, population, latitude, longitude, and elevation.

Dictionary of Physical Geography by David S. G. Thomas and Andrew Goudie. Malden, MA: Blackwell, 2000. ISBN: 0631204725.
The third edition of this comprehensive encyclopedic dictionary covers physical geography and provides an essential reference for all students and lecturers.

Merriam-Webster's Geographical Dictionary. Springfield, MA: Merriam-Webster, 1997. ISBN: 0877795460.
A listing of over 48,000 mountains, lakes, towns, and countries including pronunciation, location, and brief textual data. Older editions are also useful.

Modern Dictionary of Geography by Ronald J. Small, Simon R.J. Ross, and Michael E. Witherick. New York: Arnold; Oxford University Press, 2001. ISBN: 034080713X.
Provides a comprehensive guide to, and in many instances an explanation of the principles, concepts and terminology of modern school geography.

Oxford Dictionary of the World by David M. Munro. New York: Oxford University Press, 1995. ISBN: 0198661843.
Describes features of the physical world and notes the political landscape through statistics and data on nations, regions, peoples, languages, religions, cities, and major towns.

Web Sites

CIA World Factbook
www.cia.gov/cia/publications/factbook/
Annually profiles the economy, government, land, demographics, and national defense establishments of nations and other political entities.

GEOnet Names Server (GNS)
http://earth-info.nga.mil/gns/html/index.html
Provides access to the National Geospatial-Intelligence Agency's and the U.S. Board on Geographic Names' database of foreign geographic feature names.

National Geographic Maps and Geography
www.nationalgeographic.com/maps/
Online edition of the magazine, plus features, news, maps, and travel guides.

Perry-Castañeda Library Map Collection, University of Texas at Austin Library
www.lib.utexas.edu/maps/index.html
Online maps of current importance and general interest.

USGS Geography Information
http://geography.usgs.gov
Provides scientific information to describe and interpret America's landscape by mapping the terrain, monitoring changes over time, and analyzing how and why these changes have occurred.

Organizations

The American Geographical Society (AGS)
www.amergeog.org
Encourages research in geography and dissemination of geographic knowledge. Sponsors research projects. Operates travel and educational programs.

Association of American Geographers (AAG)
www.aag.org
Scientific and educational society founded in 1904. Members share interests in the theory, methods, and practice of geography and geographic education.

National Geographic Online
www.nationalgeographic.com
Sponsors expeditions and research in geography, natural history, archaeology, astronomy, ethnology, and oceanography; sends writers and photographers throughout the world; produces magazines, maps, books, television documentaries, films, educational media, and information services for media.

Pan American Institute of Geography and History (PAIGH)
www.ipgh.org.mx
Countries of the Americas united to promote and publicize studies in cartography, geography, geophysics, and related sciences. Offers educational courses, workshops, and seminars. Publishes 7 periodical journals.

Subject Headings

- geography
- geography—dictionaries and encyclopedias
- geography—guidebooks
- physical geography

Geology: A Research Guide

Geology is the study of the earth and of the materials and processes that shape it. This interdisciplinary subject overlaps and interacts with many other disciplines including chemistry, physics, biology, oceanography, and atmospheric sciences. Government agencies and professional earth science societies are a source of many geological publications. This guide introduces researchers to some of the basic informational sources on the topic. The terms and phrases listed in the subject headings below can be used to search for more materials in the library's catalog and research databases. If you need further assistance, please ask a librarian.

Reference Books

Glossary of Geology, 4th ed. by Julia A. Jackson and Robert L. Bates. Alexandria, VA: American Geological Institute, 1997. ISBN: 0922152349.
Many of the entries contain a syllabification guide and background information; the origin of terms; the meaning of abbreviations and acronyms common in geoscience vocabulary; the dates that many terms were first used; the meaning of certain prefixes; and the preferred term of two or more synonyms.

Encyclopedia of Field and General Geology edited by Charles W. Finkl. New York: Van Nostrand Reinhold, 1988. ISBN: 0442224990.
Provides an introduction to general field work through selected topics that illustrate specific techniques and methodologies.

Macmillan Encyclopedia of Earth Sciences edited by E. Julius Dasch. 2 vols. New York: Macmillan Reference USA, 1996. ISBN: 0028830008.
Includes entries on the Earth's processes, resources, natural hazards, and the earth in space. Also included are historical, vocational, and biographical entries.

Magill's Survey of Science, Earth Science Series by Frank N. Magill and Roger Smith. 6 vols. Pasadena, CA: Salem Press, 1990. ISBN: 0893566063.
Information on all branches of earth science, including geology, oceanography, climatology, astronomy, and environmental issues.

McGraw-Hill Dictionary of Geology and Mineralogy edited by Sybil P. Parker. New York: McGraw-Hill, 1997. ISBN: 0070524327.
Contains 7,000 terms, covering physical geology, historical geology, plate tectonics, petrology, and stratigraphy, with definitions identified by the field in which they are primarily used.

Oxford Companion to the Earth edited by Paul L. Hancock and Brian J. Skinner. New York: Oxford University Press, 2000. ISBN: 0198540396.
Includes entries on all aspects of earth science, along with black-and-white illustrations. Biographies of important scientists and explanations of key principles and terms.

Journals

Geology. Geological Society of America. 1973– . Monthly. ISSN: 00917613.

Geological Society of America Bulletin. Geological Society of America. 1961– . Monthly. ISSN: 00167606.

Geotimes. American Geological Institute. 1956– . Monthly. ISSN: 00168556.

The Journal of Geology. University of Chicago Press. 1893– . Bimonthly. ISSN: 00221376.

Web Sites

American Geological Institute (AGI)
www.agiweb.org
Nonprofit organization that provides information services to geoscientists, serves as a voice for the profession, plays a major role in strengthening geoscience education.

Atlas of Igneous and Metamorphic Rocks, Minerals and Textures
www.geolab.unc.edu/Petunia/IgMetAtlas/mainmenu.html
Part of the virtual geology Web site at the University of North Carolina.

Geosciences Libraries and Resources in the United States and Canada
www.mines.edu/library/reference/map.html
A listing of geological libraries and resources in the United States and Canada, arranged by state.

A Tapestry of Time and Terrain
http://tapestry.usgs.gov
Combined geologic and topographic map.

U.S. Geological Survey
www.usgs.gov
Federal resource for science about the earth, its natural and living resources, natural hazards and the environment.

USGS National Geologic Map Database
http://ngmdb.usgs.gov
Provides a finding tool for over 67,000 maps; over 2,000 are now available online. Researchers will also find a lexicon of geologic names of the United States.

Subject Headings

- climatology
- earth sciences
- geology
- hydrology
- mineralogy
- paleontology
- petrology
- stratigraphy

Government Information: A Research Guide

Government documents are publications issued by all levels of government, either to meet legal requirements or to provide useful information for the public. Below you will find a list of the most frequently requested federal government publications organized by topic and branch of government. In addition to the government publications below, when government reports or laws are issued, articles about them appear as secondary sources. These articles may provide summaries and explanations of the primary document and can be located by searching the library's periodical databases. The terms and phrases listed in the subject headings below can be used to search for more materials in the library's catalog and research databases. If you need further assistance, please ask a librarian.

Statistical Information

County and City Data Book. United States. Washington, DC: U.S. Department of Commerce, Bureau of the Census. 1949– . ISSN: 00829455.
www.census.gov/statab/www/ccdb.html
Including statistics for all U.S. counties, cities of 25,000 or more population, and places with more than 2,500 inhabitants. Useful for both business and school projects. Population, housing, business, and labor force data are featured.

Statistical Abstract of the United States. United States. Washington, DC: Government Printing Office, 1878– . ISSN: 00814741.
www.census.gov/statab/www/
Statistics describing social, economic, political, and geographic changes in the United States.

Uniform Crime Reports: Crime in the US
www.fbi.gov/ucr/ucr.htm
Provides detailed statistical information about crimes as reported to the FBI.

United States Census
www.census.gov
The ten-year census required by law provides massive amounts of population, housing, economic, and geographic data. The economic census, as well as other kinds of statistical data, also is found here.

World Fact Book
www.cia.gov/cia/publications/factbook/index.html
The Factbook is a comprehensive resource of facts and statistics on more than 250 countries.

Executive Branch Documents

Budget of the U.S. Government
www.gpoaccess.gov/usbudget/
Provides information on how the federal government spends money, from *A Citizens Guide to the Federal Budget*, written for the average reader, to extensive economic and historical analyses of the United States budget.

Code of Federal Regulations
www.gpoaccess.gov/cfr/index.html
The CFR contains administrative regulations, issued by federal agencies or presidential executive order, which are arranged by subject and published annually. Weekly updates are found in the Federal Register.

Federal Register
www.gpoaccess.gov/fr/index.html
Daily publication that presents final regulations as they are enacted, proposed rules, notices, and presidential documents. The same information is organized by subject and published annually in the Code of Federal Regulations.

Public Papers of the President
www.gpoaccess.gov/pubpapers/index.html
Papers and speeches of the President of the United States that were issued by the Office of the Press Secretary during the specified time period.

United States Government Manual
www.gpoaccess.gov/gmanual/index.html
Manual provides information on the agencies of the three branches of government.

Weekly Compilation of Presidential Documents
www.gpoaccess.gov/wcomp/index.html
The Weekly Compilation of Presidential Documents is published every Monday by the Office of the Federal Register, National Archives and Records Administration and contains statements, messages, and other presidential materials released by the White House during the preceding week.

Legislative Branch Documents

Congressional Directory
www.gpoaccess.gov/cdirectory/index.html
Staff directory for the legislative, executive, and judicial agencies of the federal government. It features brief biographies, committee memberships, and addresses and phone numbers for staff members and judicial branches.

Legislative Branch Resources on GPO Access
www.gpoaccess.gov/legislative.html
History of bills, Congressional bills, Congressional record, full-text versions of all signed bills, and more.

Public and Private Laws
www.gpoaccess.gov/plaws/index.html
Public and private laws are prepared and published by the Office of the Federal Register (OFR), National Archives and Records Administration (NARA). GPO Access contains the text of public and private laws enacted from the 104th Congress to the present.

United States Code
www.gpoaccess.gov/uscode/index.html
The United States Code is the permanent, or public, laws of the United States organized into broad subject areas called titles.

United States Constitution
www.gpoaccess.gov/constitution/index.html
The basic document establishing the government of the United States, the Constitution of the United States also defines the rights and liberties of citizens.

Judicial Branch Documents

Supreme Court Opinions
www.supremecourtus.gov
The Supreme Court of the United States is the final arbiter in controversies or interpretation of the laws of the United States or the intent of the Constitution. Its decisions are called "Opinions."

Subject Headings

- executive departments—United States—handbooks, manuals, etc.
- United States—Congress—joint committee on printing
- United States—Congress directories
- United States—Office of the Federal Register
- United States—politics and government—handbooks, manuals, etc.

Grants: A Research Guide

Organizations looking for money to fund various community projects must first identify prospective funding agencies whose mission statement reflects the outcome of the project. Long before this process even starts the planning phase must begin. The resources below will help researchers not only identify potential sources of funding, but instruct in the process from the initial stages of planning, the application process and reporting on a grants progress. The terms and phrases listed in the subject headings below can be used to search for more materials in the library's catalog and research databases. If you need further assistance, please ask a librarian.

Books

Annual Register of Grant Support. New Providence, NJ: R. R. Bowker, 1969– . Annual. ISSN: 00664049.
Guide to more than 3,500 grant-giving organizations offering nonrepayable support. Organized by 11 major subject areas.

Catalog of Federal Domestic Assistance. Washington, DC: The Office of Management and Budget, 1970– . Annual. ISSN: 00977799. Also available at *www.cfda.gov.*
Provides information on all federal programs available to state and local governments; federally-recognized Indian tribal governments; territories of the United States; domestic public, quasi-public, and private profit and nonprofit organizations and institutions; specialized groups; and individuals.

Directory of Research Grants 2004. Phoenix: Oryx Press, 2003. Annual. ISSN: 1573565954.
Entries arranged under broad topics. Each entry gives descriptive information such as amount of awards, purpose, and application guidelines. Indexes organized by grant names, organizations, and subjects.

The Foundation Center's Guide to Proposal Writing, 4th ed. by Jane C. Geever. New York: Foundation Center, 2004. ISBN: 1931923922.
Covers writing a grant proposal including planning stages, drafting the copy, and post-grant follow up with attention to researching, contacting, and cultivating potential sources of funds.

Foundation Directory. New York: Foundation Center, 1960– . Annual. ISSN: 00718092.
Features key facts on the nation's top 10,000 foundations by total giving.

Grant Writing For Dummies by Beverly A. Browning. Foster City, CA: IDG Books Worldwide, 2001. ISBN: 0764553070.
Provides an overview of what types of grants are available and how to apply for them, grant-writing terminology, samples of successful grant proposals.

The "How To" Grants Manual: Successful Grantseeking Techniques for Obtaining Public and Private Grants, 5th ed. by David G. Bauer. Phoenix: Oryx Press, 2003. ISBN: 0275980707.
Designed to helps readers identify best sources of funds, develop systematic approach, write tailored proposals, and learn to work well with government, foundation, and corporate funders.

Proposal Planning and Writing, 3rd ed. by Lynn E. Miner and Jeremy T. Miner. Phoenix: Oryx Press, 1998. ISBN: 1573564982.
Covers finding and developing proposals, writing foundation and corporate proposals, writing government proposals, and final steps to successful proposal writing.

Secrets of Successful Grantsmanship: A Guerrilla Guide to Raising Money by Susan L. Golden. San Francisco: Jossey-Bass, 1997. ISBN: 078790306X.

Gives fundraisers the grantsmanship skills they need to successfully navigate the grant-making process, from conducting effective prospect research and making initial conversations count to preparing, submitting, and following up on grant proposals.

Web Sites

The Foundation Center
www.fdncenter.org
Includes online databases, news, training programs, locations of learning centers, Foundation Center libraries, and cooperating collections.

Grants.gov
www.grants.gov
Allows organizations to electronically find and apply for competitive grant opportunities from all Federal grant-making agencies.

SRA International Grants Web
www.srainternational.org/newweb/grantsweb/index.cfm
Directory of public and private research funding resources from the Society for Research Administrators.

Subject Headings

- economic assistance, domestic
- endowments—directories
- foundations—directory
- fund raising
- grants-in-aid—United States
- nonprofit organizations
- proposal writing for grants—United States
- research grants
- research support

Health and Wellness: A Research Guide

Knowledge about therapies, medication, remedies, nutrition, and fitness empower patients to make better health and lifestyle decisions. Health information also allows patients to have more meaningful dialogue with their health care providers. The resources below provide a starting point for researchers as they begin to search for health information. The terms and phrases listed in the subject headings below can be used to search for more materials in the library's catalog and research databases. If you need further assistance, please ask a librarian.

Reference Books

Alternative Medicine, the Definitive Guide, 2nd ed. edited by Larry Trivieri and John W. Anderson. Berkeley: Celestial Arts, 2002. ISBN: 1587611406.

Profiles alternative therapies—acupuncture, craniosacral therapy, qigong, hyperthermia, and yoga among them—including their development, how they work, proven uses, and controversial or yet-to-be-proven uses, along with contact information for organizations.

The Alternative Medicine Handbook: The Complete Reference Guide to Alternative and Complementary Therapies by Barrie R. Cassileth. New York: W. W. Norton, 1998. ISBN: 0393045668.

Describes some of the most popular alternative therapies. Each therapy is discussed in terms of its origins, the reasons practitioners say it works, some reference to scientific evidence, and a listing of resources.

The American Medical Association Family Medical Guide, 3rd ed. edited by Charles B. Clayman. New York: Random House, 1994. ISBN: 0679412905.

Explanations of more than 650 diseases and disorders, practical advice about preventive health care, up-to-date discussion of timely health issues.

Also available on CD-ROM. New York: Dorling Kindersley, 1996. ISBN: 0789402971.

The Common Symptom Guide, 5th ed. by John H. Wasson. New York: McGraw-Hill Professional, 2002. ISBN: 0071377654.

Quick-reference handbook provides a concise listing of pertinent questions on patient and family history, physical examination, diagnostic considerations, and medications.

The Complete Home Wellness Handbook: Home Remedies, Prevention, Self-Care by John E. Swartzberg and Sheldon Margen. New York: Rebus, 2001. ISBN: 0929661656.

Complete home health encyclopedia which allows readers to self-diagnosis and treat many common minor ailments from arthritis to snoring.

Encyclopedia of Human Nutrition edited by Michèle J. Sadler, James J. Strain, and Benjamin Caballero. 3 vols. San Diego: Academic Press, 1999. ISBN: 0122266943.

Cover the scientific, political, and social aspects of nutrition including individual nutrients and foods, anatomy and physiology, nutritional therapy for diseases and conditions.

Encyclopedia of Nutrition and Good Health, 2nd ed. by Robert A. Ronzio. New York: Facts On File, 2003. ISBN: 0816049661.

Foods, their ingredients, and nutritional values are described. Specific diets (Atkins, Mediterranean) are discussed objectively, with the basic premise of the diet explained along with its pros and cons. Entries on foods and the components implicated in diseases and disorders explain how and why the problem occurs and offer dietary recommendations.

The Gale Encyclopedia of Alternative Medicine, 2nd ed. edited by Kristine M. Krapp and Jacqueline L. Longe. 4 vols. Detroit: Gale Group, 2004. ISBN: 0787674249.

Contains 750 alphabetically arranged articles covering 157 therapies, 238 diseases and conditions, and 306 herbs and other remedies. Includes illustrations, photographs, glossary and biographies of important practitioners.

The Harvard Medical School Family Health Guide. London: Cassell, 2001. ISBN: 0304357197. Also available at *www.health.harvard.edu/fhg/*
Includes chapters on lifestyle changes that will enhance health, such as nutrition, exercise, smoking cessation, safe sex, eldercare, death and dying, drug interactions, and emergency care.

Health Care Terms, 4th ed. by Vergil N. Slee, Debora A. Slee, and H. Joachim Schmidt. St. Paul, MN: Tringa Press, 2001. ISBN: 1889458023.
Definitions of nonmedical terms relating to the health care industry in the United States. Listings include, for example, professional associations, new procedures and therapies, diagnostic terms, providers, legal terms, types of coverage, and government regulations.

Mayo Clinic Guide to Self-Care: Answers for Everyday Health Problems, 3rd ed. by Philip T. Hagen. Rochester, MN: Mayo Clinic, 2003. ISBN: 1893005305.
Reliable, and easy-to-understand health information on 150 medical conditions and issues relating to health.

The PDR Family Guide to Common Ailments edited by David W. Sifton. New York: Ballantine Books, 2001. ISBN: 0345417151.
Home resource to more than 350 medical problems and procedures.

Web Sites

Healthfinder. Washington, DC: Office of Disease Prevention and Health Promotion, U.S. Department of Health and Human Services.
www.healthfinder.gov
Gateway consumer health information Web site from the United States Government.

Medical Encyclopedia (From MedlinePlus)
www.nlm.nih.gov/medlineplus/encyclopedia.html
Access to articles about diseases, tests, symptoms, injuries, and surgeries. It also contains an extensive library of medical photographs and illustrations.

PDRHealth
www.gettingwell.com
Consumer information from the *Physicians' Desk Reference* about detecting, preventing, and treating a variety of medical conditions. Includes disease overviews, drug information, and clinical trials.

WebMD
www.webmd.com
Includes articles, medical information, doctor and clinic search, health and wellness, and online tools.

Subject Headings

- alternative medicine—handbooks, manuals, etc.
- complementary therapies—handbooks
- diagnosis—popular works
- diet and nutrition
- family—health and hygiene
- fitness
- health—popular works
- medicine—popular works
- prevention
- self-care
- smoking cessation

Hispanic Heritage: A Research Guide

Hispanic-Americans, or Latinos, are a diverse group of U.S. residents defined by their common connection to the Spanish language. This growing population has a major impact on national politics, as it is the largest minority group in the United States. Their cultural contributions can be discovered by examining the following resource materials. This guide introduces the researcher to some of the basic informational sources on the topic. The terms and phrases listed in the subject headings below can be used to search for more materials in the library's catalog and research databases. If you need further assistance, please ask a librarian.

Books

Encyclopedia of Latino Popular Culture in the United States by Cordelia Candelaria, Peter J. García, and Arturo J. Aldama. 2 vols. Westport, CT: Greenwood Press, 2004. ISBN: 0313322155.
Contains 500 entries on noted people, festivities, items, terms, movements, sports, food, events, places, visual and performing arts, film, institutions, fashion, literature, organizations, the media, and more.

Hispanic Firsts: 500 years of Extraordinary Achievement by Nicolás Kanellos. Detroit: Gale Group, 1997. ISBN: 0787605174.
Chronicles 1,000 Hispanic accomplishments related to labor, religion, business and commerce, publishing, the arts, sports, science and technology, government, civil rights, and other fields.

Latinos: A Biography of the People by Earl Shorris. New York: W. W. Norton, 1992. ISBN: 0393033600.
Explores the collective lives of Mexican-Americans, Cuban-Americans, Puerto Ricans, and other Spanish-speaking descendants of the Spanish conquest of the native populations of the New World.

Notable Latino Americans: Biographical Dictionary edited by Matt S. Meier, Conchita Franco Serri, and Richard A. Garcia. Westport, CT: Greenwood Press, 1997. ISBN: 0313291055.
A biographical dictionary devoted to the accomplishments of American men and women of Latino heritage.

Museums

Mexican Fine Arts Center Museum
www.mfacmchicago.org
A Chicago museum that focuses heavily on educational programs for children. They also house a large collection of Mexican paintings and sculpture.

Mexican Heritage Plaza
www.mhcviva.org
A cultural center in San Jose that hosts art exhibits, concerts, and other events pertaining to Latin-American culture and serves as a resource center for educational programming.

Museo del Niño (in Spanish)
www.museodelninopr.org
A Puerto Rican museum devoted to arts and educational activities for children.

Museo de Arte de Ponce (in Spanish)
www.museoarteponce.org
A Puerto Rican museum devoted to classical art, with more than 3,000 paintings from the 14th to the 20th century. The site itself is a work of art and displays many of the pieces that are kept in the permanent collection.

El Museo de Barrio
www.elmuseo.org
A New York City museum devoted to Puerto Rican, Caribbean, and Latin-American art. The museum houses both contemporary and classical works of art, ranging from pre-Colombian pottery to postmodern film.

Museum of Latin American Art
www.molaa.com
A museum in Long Beach, California founded by Dr. Robert Gumbiner with the goal of educating the American public about the artistic contributions of Latin-Americans. The Web site features a virtual gallery so that you can tour the museum without leaving home.

Smithsonian Center for Latino Initiatives
http://latino.si.edu/virtualgallery/start.html
A visually stunning virtual museum of art with several interactive exhibits.

Hispanic Art Web Sites

Del Corazón
http://nmaa-ryder.si.edu/webzine/index.html
Featuring the Smithsonian American Art Museum's collection of art by Latino artists, ¡del Corazón! is an interactive, educational Webzine for teachers and students.

MuseumStuff.com
www.museumstuff.com/museums/types/hispanic_american/
A primary gateway page for Hispanic Art museums, historical societies, and arts organizations.

National Park Service
www.cr.nps.gov/nr/feature/hispanic/
Celebrate National Hispanic Heritage Month with the National Register of Historic Places.

Subject Headings

- Hispanic Americans
- Hispanic Americans—biography
- Hispanic Americans—history

Inventors and Inventions: A Research Guide

Inventions have changed the way that people live, where they live, and how they work. Many devices that you can find at school, home, or at work have been invented and at one time never existed. The creation or invention of that device solved a problem. For example, the garbage disposal was invented in 1927 by the architect and inventor John W. Hammes who built one for his wife in 1927. It took 10 years of design improvement before he was able to begin selling his appliance to the public. Hammes began the In-Sink-Erator Manufacturing Company. However small or large, inventions have changed the way that we live. The materials listed below will help get researchers started as they look for information about inventors and inventions. If you need further assistance, please ask a librarian.

Books

Inventing Modern America: From the Microwave to the Mouse by David E. Brown. Cambridge, MA: MIT Press, 2002. ISBN: 0262025086.
Profiles thirty-five inventors who exemplify the rich technological creativity of the United States over the past century including George Washington Carver, Henry Ford, and Steve Wozniak, as well as lesser known inventors such as Stephanie Kwolek, inventor of Kevlar and Wilson Greatbatch, inventor of the first implantable cardiac pacemaker.

Mothers of Invention: From the Bra to the Bomb: Forgotten Women and Their Unforgettable Ideas by Ethlie Ann Vare and Greg Ptacek. New York: Morrow, 1988. ISBN: 0688064647.
Focus on women inventors or women whose ideas were stolen by men. Covers all time periods, although many 20th-century inventors are represented.

Popular Patents: America's First Inventions from the Airplane to the Zipper by Travis Brown. Lanham, MD: Scarecrow Press, 2000. ISBN: 1578860105.
Contains more than eighty stories of America's first inventions, among them the adding machine, bottle cap, helicopter, and submarine. Each chapter includes a sketch of the invention, a profile of the inventor, and a fascinating glimpse of how that particular invention has found its way into American culture.

Web Sites

The American Experience: Technology
www.pbs.org/wgbh/amex/technology/
Profiles major technological advances in American. Also publishes a technology time line and information about forgotten inventors and their inventions.

How Stuff Works
www.howstuffworks.com
Explains in simple words how things work. Search or browse by these topics: computers, electronics, automobiles, science, home, entertainment, health, money, travel, and people.

Inventors
http://inventors.about.com
This site provides a wide range of information about famous inventors and inventions arranged in alphabetical orders, computer inventions, a time line and inventions of the industrial revolution. Also lists women, black, and kid inventors.

The Official Rube Goldberg Web Site
www.rubegoldberg.com
Best known for his invention cartoons, which use a string of outlandish tools, people, plants, and steps to accomplish everyday simple tasks in the most complicated way.

Time Magazine's Coolest Inventions
www.time.com/time/2003/inventions/
Includes the coolest inventions in the categories of music and fashion, health and safety, gadgets and robots, transportation, extreme sports, and light and dark.

Organizations

Lemelson-MIT Program
http://web.mit.edu/invent/index.html
A program that is dedicated to honoring inventors and encouraging tomorrow's great inventors. Web site Includes inventors of the week and the Inventor's Handbook, a worthwhile resource for budding inventors.

The National Inventors Hall of Fame
www.invent.org
The National Inventors Hall of Fame honors the women and men responsible for the great technological advances that make human, social, and economic progress possible. Provides brief synopsis on their lives and inventions.

United States Patent and Trademark Office
www.uspto.gov and *www.uspto.gov/go/kids/*
Provides a searchable database of inventions patented back to 1790. Kids' pages include frequently asked questions, scavenger hunts, and games.

Subject Headings

- inventions—history—18th century
- inventions—history—19th century
- inventions—United States—history—20th century
- inventors—United States—biography
- inventrices
- technology—history—18th century
- technology—history—19th century
- women inventors

Legal Information: A Research Guide

Whether researching a legal topic for a paper or in need of legal assistance, the resources below will assist all researchers with finding basic legal sources of information. From finding forms to Supreme Court rulings, the listings below will prove useful for locating information about the legal systems of the Federal government, legal education, the legal profession, and procedural and contact information.

Reference Books

Black's Law Dictionary, 7th ed. edited by Bryan A. Garner. St. Paul: West Group, 1999. ISBN: 0314241302.
Comprehensive legal dictionary.

Encyclopedia of the U.S. Supreme Court edited by Thomas T. Lewis and Richard L. Wilson. 3 vols. Pasadena, CA: Salem Press, 2001. ISBN: 0893560979.
Essays on Supreme Court decisions; historical overviews of how the Court has treated important issues; specific historical events and eras; court administration and structure; legal terms; judicial interpretation of state and federal laws; and biographies of Supreme Court justices.

Fundamentals of Legal Research, 8th ed. by Roy M. Mersky and Donald J. Dunn. New York: Foundation Press, 2002. ISBN: 158778064X.
Discussion of legal research materials and research strategy and process.

Martindale-Hubbell Law Digest (current title: *Martindale-Hubbell International Law Digest*). Summit, NJ: Martindale-Hubbell, 1991– . Annual. ISSN: 10889779
Provides a quick and concise summary of the law on various legal topics for each of the 50 states, DC, Puerto Rico and the Virgin Islands. Includes citations to statutes and other legal authority. Also includes English-language summaries of the laws of 80 countries, including new sections on Vietnam, the Slovak Republic, the Republic of Latvia, and the Island of Guernsey.

National Survey of State Laws, 4th ed. Detroit: Gale Research, 2003. ISBN: 0787656941.
Compares state laws in 46 subject areas such as business and consumer laws, prayer in public schools, marital property, and homestead.

Thomas: Legislative Information on the Internet
http://thomas.loc.gov/home/thomas2.html
Contains the full-text of U.S. public laws, bill texts, committee reports, and other legislative information maintained by the Library of Congress.

West's Encyclopedia of American Law. 12 vols. St. Paul: West., 1998. ISBN: 031405538X.
Coverage of 5,000 legal topics and biographies of people who have played a part in American law. Also includes definitions of legal terms.

Directories

BNA's Directory of State and Federal Courts, Judges and Clerks edited by Kamla J. King and Judith A. Miller. Washington, DC: Bureau of National Affairs, 1992– . Biennial. ISSN: 10785582.
State-by-state and federal listings.

Law and Legal Information Directory edited by Paul Wasserman and Marek Kaszubski. Detroit: Gale Research, 1980– . Annual. ISSN: 0740090X.
Describes a wide range of national and international organizations, services and programs related to law and the legal professional.

Legal Researcher's Desk Reference. Teaneck, NJ: Infosources, 1990– . Biennial. ISSN: 10503056.
Directory and quick reference source that includes Web and e-mail addresses.

Martindale-Hubbell Law Directory. New York: Martindale-Hubbell Law Directory, 1931– . Annual. ISSN: 01910221. Available at *www.martindale.com.*
Directory of law firms and attorneys with brief biographical information on practicing lawyers.

National Center for State Courts—Court Web Sites
www.ncsconline.org/D_KIS/info_court_web_sites.html
A directory of links to state, federal, and international court Web sites.

Official Guide to ABA Approved Law Schools (merger of *Official Guide to U.S. Law Schools* and *Official American Bar Association Guide to Approved Law Schools*). Newtown, PA: ABA-LSAC, 2002– . Annual. ISSN: 15343502. Available at *http://officialguide.lsac.org/docs/cgi-bin/home.asp*
Includes statistics such as enrollment, tuition and expenses, faculty, curriculum and bar passages rates. Law school Web addresses are also provided.

Want's Federal-State Court Directory. Washington, DC: WANT, 1984– . Annual. ISSN: 07421095.
A directory including information on the organization, structure, and jurisdiction of Federal, State, and U.S. territory courts.

West Legal Directory
http://lawyers.findlaw.com
The West Legal Directory on FindLaw's "Find a Lawyer" page allows users to search for attorneys and law firms in the United States and abroad, by name, geographic location and area/type of practice.

Forms

Court Rules, Forms, and Dockets
www.llrx.com/courtrules/
Over 1,400 sources for state and federal court rules, forms, and dockets. You can browse to find the resource that you need or search by keyword.

Findlaw Court Forms
http://forms.lp.findlaw.com
Includes court forms for federal appellate, district, and bankruptcy courts. The state forms collection contains court forms for state courts.

Legal Search Engines and Directories

FindLaw
www.findlaw.com
Indexes of Internet legal resources.

Legal Information Institute
www.law.cornell.edu
Developed by Cornell Law School, this Web site includes links to resources in the field of law.

Subject Headings

- courts—United States—directories
- courts—United States—states—directories
- judges—United States—directories
- law—United States
- lawyers—United States—directories
- legal research—United States—handbooks, manuals, etc.

Linguistics: A Research Guide

Linguistics is the scientific study of human language, including phonology, syntax, semantics, pragmatics, bilingualism, sociolinguistics, and computational linguistics. This guide introduces researchers to some of the basic informational sources on the topic. The terms and phrases listed in the subject headings below can be used to search for more materials in the library's catalog and research databases. If you need further assistance, please ask a librarian.

Books

Cambridge Encyclopedia of Language, 2nd ed. edited by David Crystal. New York: Cambridge University Press, 1997. ISBN: 0521596556.
Each section is a self-contained presentation of a major theme in language study, with cross-references included to related sections and topics. Three indexes provide access by language, language families, dialects, scripts, authors, topics, and personalities.

Compendium of the World's Languages, 2nd ed. edited by George L. Campbell. 2 vols. New York: Routledge, 2000. ISBN: 0415202981.
Contains over 5,000 separate languages, many with numerous dialects, gives brief history of each language, provides phonology, morphology, syntax, and typical word order for each language.

Encyclopedia of Language and Linguistics edited by R. E. Asher and J. M. Y. Simpson. 10 vols. New York: Pergamon, 1994. ISBN: 0080359434.
Covers many different views and perspectives encountered in research and thinking in the fields of linguistics and language studies, emphasizing the multidisciplinary nature of the subject.

International Encyclopedia of Linguistics, 2nd ed. by William J. Frawley. 4 vols. New York: Oxford University Press, 2003. ISBN: 0195139771.
Articles include bibliographies, tables, charts, and maps to illustrate concepts. Comprehensive and authoritative with special attention paid to the interrelations within branches of linguistics.

Linguistics: A Guide to the Reference Literature by Anna L. DeMiller. Englewood, CO: Libraries Unlimited, 2000. ISBN: 1563086190.
Coverage spans from 1957 to the present. Detailed citations describe and evaluate each work, often offering comparisons to similar titles.

Longman Dictionary of Applied Linguistics. Essex, England: Longman, 1985. ISBN: 0582557089.
Lengthy definitions that provide references for those who want to look deeper into a topic. The dictionary also provides cross-references to other relevant definitions.

Oxford Companion to the English Language by Tom McArthur. New York: Oxford University Press, 1998. ISBN: 0192800612.
Entries encompass writing and speech, linguistics, rhetoric, literary terms; related topics such as bilingual education, child language acquisition, sign language, and psycholinguistics; and biographies of figures who have influenced the shape or study of English.

Oxford English Dictionary, 2nd ed. edited by John A. Simpson and Edmund S. Weiner. 20 vols. New York: Oxford University Press, 1989. ISBN: 0198611862.
OED covers words from across the English-speaking world, from North America to South Africa, from Australia and New Zealand to the Caribbean. It also offers the best in etymological analysis, listings of variant spellings, and shows pronunciation using the International Phonetic Alphabet.

Roget's II: The New Thesaurus, 3rd. ed. Boston: Houghton Mifflin, 2003. ISBN: 0618254145. Available online at *www.bartleby.com/62/.*
Features of the thesaurus include succinct word definitions and an innovative hypertext linked category index, which leads you to list of antonyms for the word you select.

Routledge Dictionary of Language and Linguistics edited by Hadumod Bussuman, et al. New York: Routledge, 1996. ISBN: 0415022258.
Provides a survey of the key terminology and languages of more than 30 subdisciplines of linguistics.

Web Sites

Electronic Journals for Linguistics—University of Delaware Library
www2.lib.udel.edu/subj/ling/ej.htm
List of links to journals for linguistics.

Linguist List
www.linguistlist.org
Provides information on language and language analysis, and the discipline of linguistics with the infrastructure necessary to function in the digital world.

Linguistics, Natural Language, and Computational Linguistics Meta-index
www-nlp.stanford.edu/links/linguistics.html
Annotated list to linguistic resources on the Internet.

Linguistic Society of America
www.lsadc.org
Organization founded to advance the scientific study of language.

Subject Headings

- applied linguistics—dictionaries
- language and languages—dictionaries
- language and languages—encyclopedias
- language and languages—study and teaching—dictionaries
- linguistics—dictionaries
- linguistics—encyclopedias

LITERARY CRITICISM: A RESEARCH GUIDE

Students of literature will find the materials below to help them as they begin their research. Whether looking for biographical information and writing style of an author or critiques of certain titles, the basic resources listed below will help students find them. The terms and phrases listed in the subject headings below can be used to search for more materials in the library's catalog and research databases. If you need further assistance, please ask a librarian.

REFERENCE BOOKS

The Concise Oxford Dictionary of Literary Terms by Chris Baldick. New York: Oxford University Press, 1990. ISBN: 0198117337.
Over 1,000 literary terms including extensive coverage of traditional drama, rhetoric, literary history, and textual criticism.

Contemporary Literary Criticism: Excerpts from Criticism of the Works of Today's Novelists, Poets, Playwrights, Short Story Writers, Scriptwriters, and Other Creative Writers. Detroit: Gale Research, 1973– . Annual.
Profiles approximately six to eight novelists, poets, playwrights, and other creative writers by providing full-text or excerpted criticism taken from books, magazines, literary reviews, newspapers, and scholarly journals.

Dictionary of Literary Terms by Harry Shaw. New York: McGraw-Hill, 1972. ISBN: 0070564906.
Explains and illustrates a number of technical terms pertaining to literature, as well as films, newspapers, and magazines.

A Handbook to Literature, 8th ed. by William Harmon, et al. Upper Saddle River, NJ: Prentice Hall, 2000. ISBN: 0130979988.
Includes over 2,000 terms, including those from computing and information management as well as film, radio, TV, printing, linguistics, music, graphic arts, and classical studies.

The HarperCollins Reader's Encyclopedia of American Literature, 2nd ed. edited by George Perkins, et al. HarperCollins, 2002. ISBN: 006019815X.
Revised and updated edition of *Benet's Reader's Encyclopedia of American Literature.* Includes thousands of entries, contributed by more than 130 scholars, include biographies of novelists, playwrights, poets and critics, summaries of books and plays, descriptions of characters, definitions of literary terms and movements, and much more.

Nineteenth-Century Literature Criticism: Excerpts from Criticism of the Works of Nineteenth-Century Novelists, Poets, Playwrights, Short-Story Writers, & Other Creative Writers. Detroit: Gale Research, 1981– . Annual.
Profiles approximately four to eight literary figures who died between 1800 and 1899 by providing full-text or excerpted criticism taken from books, magazines, literary reviews, newspapers and scholarly journals.

Twentieth-Century Literary Criticism. Detroit, Gale Research, 1978– . Annual.
Excerpts from the best criticism on the major literary figures and nonfiction writers, including novelists, poets, playwrights, and literary theorists, of 1900 to 1999—the era most frequently studied in high schools.

WEB SITES

Critical Reading, A Guide
www.brocku.ca/english/jlye/criticalreading.html
An instructional guide on how to analyze and interpret literature, particularly poetry and fiction.

Literary Criticism, Internet Public Library
www.ipl.org/div/litcrit/
Contains critical and biographical Web sites about authors and their works that can be browsed by author, by title, or by nationality and literary period.

PAL: Perspectives in American Literature, A Research and Reference Guide
www.csustan.edu/english/reuben/pal/table.html
Covers major perspectives or literary movements in American literature. Discusses over 300 years of writing in America.

LitLinks
www.smpcollege.com/litlinks/home.htm
LitLinks are organized alphabetically by author within five genres: fiction, essays, poetry, drama, and critical theory.

Sparknotes
www.sparknotes.com
Offers free study guides for many literary works. Registration required.

Today in Literature
www.todayinliterature.com
Provides information on classic and international authors and poets. Includes life stories, selected links, and suggested readings.

Subject Headings

- criticism—terminology
- English language—terms and phrases
- literary form—terminology
- literature—terminology

Military Science: A Research Guide

Military science encompasses many aspects such as military art and science, supplies, equipment, military history, military-civil relations, and arms control. The resources below can assist researchers and students in learning more about military history and current issues in military life. This guide introduces researchers to some of the basic informational sources on the topic. The terms and phrases listed in the subject headings below can be used to search for more materials in the library's catalog and research databases. If you need further assistance, please ask a librarian.

Books

American Military Leaders from Colonial Times to the Present by John C. Fredriksen. 2 vols. Santa Barbara, CA: ABC-CLIO, 1999. ISBN: 1576070018.
Prominent men and women of the military are the scope of this reference work. Biographies range from two to three pages, concluding with a bibliography. Photographs and illustrations are included, and both a subject index and a list of leaders organized by their military titles can be found at the end of volume two.

Dictionary of American Military Biography edited Roger J. Spiller and Joseph G. Dawson. 3 vols. Westport, CT: Greenwood Press, 1984. ISBN: 0313214336.
Lengthy biographical essays on some 400 individuals important in U.S. military history.

Dictionary of Military Terms: A Guide to the Language of Warfare and Military Institutions, 2nd ed. edited by Trevor Nevitt Dupay, et al. New York: H. W. Wilson, 2003. ISBN: 0824210255.
All aspects of military affairs are covered: strategy, tactics, fortifications, weapons, ranks, organization, and administration.

Dictionary of Modern War edited by Edward N. Luttwak and Stuart L. Koehl. New York: HarperCollins, 1991. ISBN: 0062700219.
Explains general concepts, major organizations, weapons and weapon systems, military concepts, basic military technologies, modes of warfare, and negotiations and treaties.

Guide to Military Installations, 5th ed. edited by Dan Cragg. Mechanicsburg, PA: Stackpole Books, 1997. ISBN: 0811724840.
Provides description of American military bases, both overseas and in the United States.

International Military and Defense Encyclopedia edited by Trevor Nevitt Dupay. 6 vols. Washington, DC: Brassey's, 1993. ISBN: 0028810112.
Definitive, comprehensive, multidisciplinary, and multicultural encyclopedia of military and defense information. Focuses primarily on post-WWII events and developments.

The Military Balance. London: Institute for Strategic Studies, 1964. Annual. ISSN: 04597222.
www3.oup.co.uk/milbal/
Statistical assessment of military forces and defense expenditures.

Oxford Companion to Military History edited by Richard Holmes. Oxford: Oxford University Press. 2001. ISBN: 0198662092.
Surveys all military services from ancient to modern time periods, but emphasizes land warfare in Europe and North America from the 18th–20th centuries.

Penguin Encyclopedia of Modern Warfare: From 1850 to the Present Day edited by Kenneth Macksey and William Woodhouse. New York: Penguin, 1993. ISBN: 0140513019.
Gives quick access to personalities, battles, campaigns, strategy, tactics, intelligence handling, logistics, and technology.

Strategic Survey. London: Institute for Strategic Studies, 1966. Annual. ISSN: 04597230.
Supplements *The Military Balance*, giving narrative summaries of the strategic situation in countries and regions of the world.

Webster's American Military Biographies edited by Robert McHenry. New York: Dover, 1984. ISBN: 0486247589.
More than 1,000 biographies of persons important to the military history of the nation.

Abstracts and Indexes

Air University Library Index to Military Periodicals. Maxwell Air Force Base, AL: Air University Library. 1990–present. Quarterly Publication. ISSN: 00022586.
www.dtic.mil/search97doc/aulimp/main.htm
Indexes articles, news items, and editorials in approximately 80 English language military and aeronautical periodicals.

Magazines

Air & Space Power Journal
www.airpower.maxwell.af.mil
U.S. Air Force's primary forum for professional discourse on air and space power. Archived back to 1987.

Airman
www.af.mil/news/airman/indxflas.html
Magazine for America's airmen. Archived back to September 1995.

Internet Resources

DefenseLINK
www.defenselink.mil
Provides direct access to the information services established by the military departments and selected Department of Defense organizations.

DOD Dictionary of Military Terms
www.dtic.mil/doctrine/jel/doddict/
Contains both military terms, acronyms, and abbreviations.

Subject Headings

- armed forces
- military art and science
- military history
- military science
- national security
- naval art and science

Moving and Relocation: A Research Guide

Whether it's because of the lure of a new job, new life, or a new city, Americans are constantly moving. The following resources will help you select a new home or land a job and find schools in the city or town of your choice. These books and Web sites include information such as the cost of living, salary levels, residential area, education and employment opportunities. This guide introduces researchers to some of the basic informational sources on the topic. The terms and phrases listed in the subject headings below can be used to search for more materials in the library's catalog and research databases. If you need further assistance, please ask a librarian.

Where to Live

America's Top Rated Cities. 4 vols. Boca Raton, FL: Universal Reference, 2002. ISBN: 1930956444.
Includes a social, business, economic, demographic, and environmental profile of each city.

Chamber of Commerce
www.chamberofcommerce.com
Creates links to city chambers throughout the country in addition to state chambers of commerce, state boards of tourism, U.S. convention and visitors bureaus, American chambers abroad, and U.S. embassies.

Cities Ranked and Rated: More than 400 Metropolitan Areas Evaluated in the United States and Canada. Indianapolis: Wiley, 2004. ISBN: 076452562X.
Each city is ranked on a number of essential factors, many of which are of vital interest in today's economy. Categories include: economy and jobs, cost of living, climate, education, health and health care, crime, transportation, leisure, and arts and culture.

Homefair.com
www.homefair.com
Includes a variety of tools for people relocating including "The Relocation Wizard" to help you plan your move and the "Salary Calculator" that compares the cost of living in hundreds of U.S. cities. Also includes free school and city reports.

Making Your Move to One of America's Best Small Towns by Norman Compton. New York: M. Evans, 2002. ISBN: 0871319888.
Includes information for each of the 120 towns listed, including geography, climate, schools, recreation facilities, sales tax rates, per capita income, the cost of electricity, and the average cost of a home.

Money's Best Places to Live
http://money.cnn.com/best/bplive/
Lists America's best places to live. Search on housing costs and weather.

Moving and Relocation Sourcebook. Detroit: Omnigraphics, 2001. ISBN: 0780804317.
Features mailing addresses, local and toll-free telephone numbers, fax numbers, Web site addresses, and e-mail addresses for chambers of commerce, government offices, libraries, and other local information resources (including online resources).

New Rating Guide to Life in America's Small Cities by Kevin Heubusch and G. Scott Thomas. Amherst, NY: Prometheus Books, 1997. ISBN: 157392170X.
Provides statistics and point rating system for small cities.

Places Rated Almanac by David Savageau and Ralph D'Agostino. New York: Hungry Minds, 2000. ISBN: 0028634470.
Metropolitan areas are rated in nine categories: costs of living, job outlook, transportation, education, health care, crime, the arts, recreation, and climate.

Finding a Home

HomeBuilder.com
www.homebuilder.com
Search for new home listings, find a custom builder, planned communities, and factory-built homes.

Homescape.com
www.homescape.com
Homescape.com is a mega-portal for the following Web sites: apartments.com, homefinder.com, MovingCenter.com, and NewHomeNetwork.com.

Owner.com
www.owners.com
Completely searchable national database of homes for sale by owner.

Realtor.com
www.realtor.com
Search for homes by state, zip code, or MLS information.

Yahoo! Real Estate
http://realestate.yahoo.com
Offers a wide variety of real estate information: buying, selling, renting, financing, and improving your home.

The Move

Monstermoving.com
www.monstermoving.com
Offers a one-stop shopping site to help you manage your move.

MoversNet at the U.S. Postal Service
www.usps.com/moversnet/
Provides information on your change of address form but also information on motor vehicle and voter registration, maps, and tips on moving with kids and pets.

Apartment Locators

Apartmentguide.com
www.apartmentguide.com
Highlights: Search by geographic area and specifications. Provides photos, floor plans, and contact information. Based on the nationally distributed print publication.

Apartments.com
www.apartments.com
Search for apartments by state and city and specifications such as price range and number of bedrooms. Provides photographs and virtual tours.

Rent.net
www.rent.net
Lists apartments and rentals such as senior living, vacation property, and temporary housing.

Subject Headings

- cities and towns—ratings—United States
- family life surveys
- metropolitan areas
- moving household
- quality of life
- retirement, places of
- social indicators
- suburbs
- urban-rural migration—United States

Multicultural Literature: A Research Guide

This guide contains tips and strategies for locating biographical and critical information on American writers by ethnicity, as well as a bibliography of resources, both in print and online. This guide introduces researchers to some of the basic informational sources on the topic. The terms and phrases listed in the subject headings below can be used to search for more materials in the library's catalog and research databases. If you need further assistance, please ask a librarian.

General Reference Resources

American Ethnic Literatures: Native American, African American, Chicano/Latino, and Asian American Writers and their Backgrounds: An Annotated Bibliography by David R. Peck. Pasadena: Salem Press, 1992. ISBN: 0893566845.
Annotated bibliography of primary and secondary works related to the four groups noted in the title, including background material such as brief narrative histories.

Minority Literatures (Voice of the Shuttle)
http://vos.ucsb.edu/browse.asp?id=2746
Links to resources in minority literature.

New Immigrant Literatures in the United States. A Sourcebook to Our Multicultural Literary Heritage by Alpana S. Knippling. Westport, CT: Greenwood Press, 1996. ISBN: 0313289689.
Presents a critical introduction to post-World War II immigrant literatures of the United States, including work by Asian-American, Caribbean-American, European-American, and Mexican-American writers.

African-American

African American Writers, 2nd ed. edited by Valerie Smith. 2 vols. New York: Charles Scribner's Sons, 2001. ISBN: 068480638X.
Introduction to African-American writers who have made a significant contribution to American and world letters.

African-American Women Writers of the 19th Century
http://digital.nypl.org/schomburg/writers_aa19/
A digital collection of 52 published works by 19th-century black women writers providing access to the thought, perspectives and creative abilities of black women, as captured in books and pamphlets published prior to 1920.

Asian-American

Asian American Literature: An Annotated Bibliography by King-Kok Cheung and Stan Yogi. New York: Modern Language Association of America, 1988. ISBN: 087352960X.
Covers writers of Asian descent living in the United States or Canada, and includes listings for both primary and secondary sources.

The Asian Pacific American Heritage: a Companion to Literature and Arts edited by George Leonard. New York: Garland, 1999. ISBN: 0815329806.
Collection of articles on the work of contemporary Asian-Pacific American writers and artists.

Latino/Chicano/Hispanic-American

Biographical Dictionary of Hispanic Literature in the United States: the Literature of Puerto Ricans, Cuban Americans, and other Hispanic Writers by Nicolás Kanellos. New York: Greenwood Press, 1989. ISBN: 0313244650.
Guide of Hispanic writers living and writing in the United States and Puerto Rico.

Chicano Literature: A Reference Guide by Julio A. Martínez and Francisco A. Lomelí. Westport, CT: Greenwood Press, 1985. ISBN: 0313236917.
Entries on major Chicano authors and important topics in the study of Chicano literature. Each author entry includes brief biographic information, a discussion of major works, a bibliography of the author's writings, and sources of criticism.

Handbook of Hispanic Cultures in the United States: Literature and Art edited by Nicolás Kanellos and Claudio Esteva-Fabregat. 4 vols. Houston: Arte Publico Press, 1993–1994. ISBN: 1558850740.
Surveys the history and development of Puerto Rican, Cuban, Chicano, and Hispanic literatures in the United States.

HAPI (Hispanic American Periodicals Index)
http://hapi.gseis.ucla.edu
Contains information about Central and South America, Mexico, the Caribbean basin, the United States-Mexico border region and Hispanics in the United States.

Masterpieces of Latino Literature edited by Frank Northen Magill. New York: HarperCollins, 1994. ISBN: 0062701061.
Includes articles on classic and newly popular works of fiction and nonfiction; general essays about the plays, short stories, poetry, and essays of major Latino writers and thinkers.

Native American

Native American Authors
www.ipl.org/div/natam/
Provides brief information about Native North American authors including bibliographies of their published works, biographical information, and links to online resources including interviews, online texts, and tribal Web sites.

Native American Literatures: An Encyclopedia of Works, Characters, Authors, and Themes by Kathy J. Whitson. Santa Barbara, CA: ABC-CLIO, 1999. ISBN: 0874369320.
Provides basic information about the authors, works, and themes integral to Native American letters.

Native North American Literature: Biographical and Critical Information on Native Writers and Orators from the United States and Canada from Historical Times to the Present edited by Janet Witalec, Jeffery Chapman, and Christopher Giroux. Detroit: Gale Research, 1994. ISBN: 0810398982.
Organized in two parts—oral literature and written literature. Includes biographical sketches, brief essays evaluating each author's work, and bibliographies.

Subject Headings

- African Americans in literature
- American literature—minority authors
- Asian Americans in literature
- ethnic groups in literature
- Hispanic Americans in literature
- Indians in literature

Music: A Research Guide

Music takes many forms around the world including Western music, which is the music of Europe and the Americas. There are two chief kinds of Western music, *classical* and *popular*. Classical music includes symphonies, operas, and ballets. Popular music includes country music, folk music, jazz, and rock music. This guide introduces researchers to some of the basic informational sources on the topic and aims to assist students and researcher in learning more about the instruments that make music and the performers who interpret it. The terms and phrases listed in the subject headings below can be used to search for more materials in the library's catalog and research databases. If you need further assistance, please ask a librarian.

Dictionaries and Encyclopedias

Baker's Biographical Dictionary of Musicians, 8th ed. by Nicolas Slonimsky. New York: Schirmer Books, 1992. ISBN: 0028724151.
Contains more than 15,000 biographies ranging from several lines to eight pages in length, *Baker's* provides unparalleled coverage of the world's greatest and lesser-known musicians in the uniquely insightful and always meticulous style of lexicographer extraordinaire Nicolas Slonimsky.

The Garland Encyclopedia of World Music edited by Bruno Nettl, et al. 10 vols. New York: Garland, 1998–2002.
10-volume series that takes a cultural approach to its focus on the music of all the world's peoples. Each volume is arranged topically, regionally, or by ethnic group, and complemented by an extensive index.

International Dictionary of Black Composers edited by Samuel A. Floyd. 2 vols. Chicago: Fitzroy Dearborn, 1999. ISBN: 1884964273.
The 185 signed essays, half of which are on classical composers, are accompanied by full-page portraits, are substantial in their coverage and analysis. Lists of compositions by genre, discographies, and printed works are included.

The New Grove Dictionary of American Music edited by H. Wiley Hitchcock and Stanley Sadie. 4 vols. New York: Grove's Dictionaries of Music, 1986. ISBN: 0943818362.
Covers American music including popular, jazz, folk, regional music, American composers and performers, publishing, periodicals, libraries, and American cities.

The New Grove Dictionary of Music and Musicians, 2nd ed. by Stanley Sadie and John Tyrrell. New York: Grove, 2001. ISBN: 1561592390.
Covers the history and development of music, composers, performers, instruments, musical forms, terms and definitions, and musical cites and institutions, the set covers every type of music, from classical to popular to jazz. The edition contains more than 29,000 articles from more than 6,000 contributors and 5,000 illustrations.

Popular Musicians edited by Steve Hochman. 4 vols. Pasadena, CA: Salem Press, 1999. ISBN: 0893569860.
Includes over 500 alphabetically arranged, signed articles recap the careers, successes, and critical reputations of a wide range of contemporary artists. Also includes discography, awards, and cross-references to other relevant articles, glossary, bibliography, time line of first releases, index of song and album titles, and list of articles by musical style.

Song Indexes

The Music Index. Warren, MI: Harmonie Park, 1949– . Quarterly.
An index to musical articles in over 300 periodicals that covers biographical, critical, historical, and analytical articles; notes premieres, obituaries, contests and awards, discography, and iconography. Arranged by subject.

Popular Song Index by Patricia Havlice. Metuchen, NJ: Scarecrow Press, 1975. ISBN: 081080820X; First supplement, 1978. ISBN: 0810810999; Second supplement, 1984. ISBN: 0810816423; Third supplement, 1989. ISBN: 0810822024; Fourth supplement, 2005. ISBN: 0810852608.

Includes 35,000 songs dating back to 1940 from 710 sources including pop, folk, blues, gospel, art and opera. Indexed by title, 1st line, composer, and lyricist.

Song Finder: A Title Index to 32,000 Popular Songs in Collections, 1854–1992 edited by Gary Lynn Ferguson. Westport, CT: Greenwood Press, 1995. ISBN: 0313294704.

Song Index: An Index to More than 12,000 Songs in 177 Song Collections Comprising 262 Volumes and Supplement edited by Minnie Earl Sears and Phyllis Crawford. 2 vols. North Haven, CT : Shoe String Press, 1966.

Web Sites

8 Notes
www.8notes.com
Free classical and traditional sheet music, free popular and jazz riffs, free music lessons, free music resources.

All Music
www.allmusic.com
Guide to music including information on all types of music including rock, country, jazz, rap, reggae, new age, gospel, and easy listening. Search by artists, albums, songs, styles, labels.

Essentials of Music
www.essentialsofmusic.com
Guide to classical music including eras, composers, and a glossary.

Music for the Nation: American Sheet Music, 1820–1860 and 1870–1885
http://memory.loc.gov/ammem/mussmhtml/
Contains more than 62,500 pieces of historical sheet music registered for copyright: more than 15,000 registered during the years 1820–1860 and more than 47,000 registered during the years 1870–1885.

Subject Headings

- music—20th century—bio-bibliography—dictionaries
- music—bio-bibliography
- music—encyclopedias
- music—United States—bio-bibliography
- music—United States—encyclopedias
- popular music—encyclopedias
- songs—19th century—indexes
- songs—20th century—indexes
- songs—indexes
- world music—encyclopedias

Mythology: A Research Guide

In early times, when people did not have the knowledge to explain certain things, they developed myths that usually involved gods or divine beings that usually had supernatural powers. These myths, often having religious significance, told us a lot about people and the way they lived. The materials listed below provide beginning researchers with a starting point for finding information on gods and deities, myths and legends, and mythology from many different places and times. The terms and phrases listed in the subject headings below can be used to search for more materials in the library's catalog and research databases. If you need further assistance, please ask a librarian.

Books and General Reference Resources

Dictionary of World Mythology by Arthur Cotterell. New York: Oxford University Press, 1997. ISBN: 0192177478. Available at www.oxfordreference.com.
Presents the gods of Greece, Rome, and Scandinavia, the mystical deities of Buddhist and Hindu India, and the spirits of the African and American continents.

Facts On File Encyclopedia of World Mythology and Legend, 2nd ed. edited by Anthony S. Mercatante and James R. Dow. 2 vols. New York: Facts On File, 2004. ISBN: 0816047081.
Illustrated guide to who's who in world mythology.

The Golden Hoard: Myths and Legends of the World by Geraldine McCaughrean and Bee Willey. New York: M. K. McElderry, 1996. ISBN: 0689807414.
Features myths, legends, and folktales from many different cultures and ethnicities.

Greek and Roman Mythology A to Z: Young Reader's Companion, rev. ed. by Kathleen N. Daly and Marian Rengel. New York: Facts On File, 2003. ISBN: 0816051550.
Alphabetically listed entries identify and explain the characters, events, important places, and other aspects of Greek and Roman mythology.

The Illustrated Book of Myths: Tales and Legends of the World by Neil Philip and Nilesh Mistry. Boston: Dorling Kindersley, 1995. ISBN: 0789402025.
Myths from all over the world, including a selection of the great classical tales of Greece and Rome and myths from Norse, Celtic, Egyptian, Native American, Aboriginal, African, and Asian traditions.

The Illustrated Bulfinch's Mythology by Thomas Bulfinch and Giovanni Caselli. New York: Macmillan USA, 1997. ISBN: 0028614755. Original text available at *www.bulfinch.org*.
Illustrated edition of classic tales of the gods and goddesses, Greek and Roman antiquity; Scandinavian, Celtic, and Oriental fables and myths.

The Illustrated Dictionary of Greek and Roman Mythology by Michael Stapleton. New York: P. Bedrick Books, 1986. ISBN: 0872260631.
Illustrated stories of gods such as Zeus, Hera, Aphrodite, Athena, and their Roman counterparts.

Mythology by Edith Hamilton and Steele Savage. Boston: Back Bay Books, 1998. ISBN: 0316341517.
A collection of Greek and Roman myths arranged in sections on the gods and early heroes, love and adventure stories, the Trojan war, and a brief section on Norse mythology.

Myths of the Hindus & Buddhists by Ananda K. Coomaraswamy and Sister Nivedita. Mumbai; Delhi: Jaico House, 1999. ISBN: 8172248210.
Indian myths retold include almost all of those, which are commonly illustrated in Indian sculpture and painting.

Who's Who in Classical Mythology by Michael Grant and John Hazel. New York: Routledge, 2002. ISBN: 0415260418.
Guide to all the Greek and Roman mythological characters, from major deities such as Athena and Bacchus to the lesser-known wood nymphs and centaurs. Includes such heroic mortals as Jason, Aeneas, Helen, Achilles, and Odysseus.

Who's Who in Non-Classical Mythology by E. Sykes and A. Kendall. London: Routledge, 2002. ISBN: 041526040X.
Presents accounts of the gods, heroes, and myths of Europe (including the Celts, Teutons, Slavs, and Basques), the Americas, Africa, Australia, China, Japan, and Indonesia.

Web Sites

Bulfinch's Mythology: The Age of Fable
www.bulfinch.org
The complete original text with information about the author.

Encyclopedia Mythica
www.pantheon.org
Encyclopedia on mythology, folklore and legends, divided into several areas including mythology, folklore, bestiary, heroes, image gallery, and genealogy tables.

Godchecker Mythology Encyclopedia
www.godchecker.com
Currently features over 1,600 deities including African, Australian, Aztec, Chinese, Egyptian, Finnish, Greek, Incan, Mayan, Native American, and Norse Gods.

Mythography
www.loggia.com/myth/myth.html
Presents resources and reference materials about Greek, Roman, and Celtic mythology.

Myths and Legends
www.myths.com/pub/myths/myth.html
Annotated list of Web sites arranged by geography.

Subject Headings

- animals, mythical
- goddesses, Greek
- gods, Greek
- legends
- myth
- mythology, Greek

Native American Heritage: A Research Guide

Learn more about the culture, way of life, and history of Native Americans by researching their foods, art forms, music, languages and spiritual beliefs of Native American people by using the materials listed below. This guide introduces researchers to some of the basic informational sources on the topic. The terms and phrases listed in the subject headings below can be used to search for more materials in the library's catalog and research databases. If you need further assistance, please ask a librarian.

Books

Atlas of the North American Indian by Carl Waldman and Molly Braun. New York: Facts On File, 1985. ISBN: 0871968509.
Covers the entire history, culture, and tribal locations of the Indian peoples of the United States, Canada, and Central America, from prehistoric times to the present day. Over 100 two-color maps.

Gale Encyclopedia of Native American Tribes edited by Sharon Malinowski and Anna Sheets. 4 vols. Detroit: Gale Group, 1998. ISBN: 0787610852.
Covers almost 400 North American tribes including historic and current location, population data, history, religious beliefs, language, buildings, means of subsistence, clothing, healing practices, customs, oral literature, and current tribal issues.

Handbook of North American Indians by William C. Sturtevant. Washington, DC: Smithsonian Institution, 1900– . Multivolume set.
Encyclopedia set provides topical and geographical coverage of North American Indians. Authoritative source in this field. Each volume covers a different geographic area.

Native American Literatures: An Encyclopedia of Works, Characters, Authors and Themes by Kathy J. Whitson. Santa Barbara, CA: ABC-CLIO, 1999. ISBN: 0874369320.
Lists more than 300 alphabetically arranged entries including authors, works, characters in works, and terms or events of historical importance.

Web Sites

AIROS
www.airos.org
Listen live now! AIROS, American Indian Radio On Satellite, represents Native America in the public broadcasting system.

Celebrating Hispanic Heritage
www.galegroup.com/free_resources/chh/index.htm
Includes biographies, culture quiz, time line, holidays, and musical genres.

Code Talk (U.S. Department of Housing and Urban Development, Office of Native American Programs)
www.codetalk.fed.us
Provides Information from government agencies and other organizations to Native American communities.

National American Indian Heritage Month
www.cr.nps.gov/nr/feature/indian/
Showcases historic properties listed in the National Register, National Register publications, and National Park units of importance to American Indians and Alaska Natives cultures.

National Native News
www.nativenews.net
Examines the social, economic, and cultural issues of both Native people and their non-Native neighbors.

Native American Authors
www.ipl.org/div/natam/
Provides information on contemporary Native North American authors with bibliographies of their published works, biographical information, and links to online resources including interviews, online texts and tribal Web sites.

Native American Business Alliance
www.native-american-bus.org
A resource of and for Native American businesses.

Native Youth Alliance
www.nativeyouthalliance.org/nya/home.asp
An advocate organization dedicated to ensuring the continuation of traditional and spiritual culture in Native American.

Notable American Indians
www.infoplease.com/spot/aihmbioaz.html
Contains brief biographical information on Native American athletes, writers, entertainers, chiefs, activists, and religious leaders.

Organizations

American Indian Movement
www.aimovement.org
Encourages self-determination among American Indians and was created to establish international recognition of American Indian treaty rights.

Association on American Indian Affairs (AAIA)
www.indian-affairs.org
Provides legal and technical assistance to Indian tribes throughout the United States in health, education, economic development, resource utilization, family defense, and the administration of justice.

Indian Heritage Council (IHC), Henry St. Box 2302, Morristown, TN 37816 USA
Promotes and supports American Indian endeavors. Seeks a deeper understanding between American Indians and others of the cultural, educational, spiritual, and historical aspects of Native Americans. Conducts research and educational programs.

Native American Indian Chamber of Commerce of North America
www.usaindianinfo.org
Clearinghouse of information on American Indians. Collects and disseminates information at major American Indian events, conferences, and gatherings. Sponsors educational programs, pow wows, and Indian art festivals.

Subject Headings

- Indians of North America—cultural assimilation
- Indians of North America—culture
- Indians of North America—diseases
- Indians of North America—domestic life
- Indians of North America—social conditions
- Indians of North America—social life
- Indians of North America—(state)

Natural Disasters: A Research Guide

Natural disasters such as drought, earthquakes, hurricanes, lightning, tornadoes, tsunamis, and volcanoes occur almost everyday somewhere on Earth. Science students and researchers will find a great many resources listed below to help them understand why and how these disasters occur and how to prepare for them. Researchers will also learn about the impact that natural disasters have on the people and places in which they occur. This guide introduces researchers to some of the basic informational sources on the topic. The terms and phrases listed in the subject headings below can be used to search for more materials in the library's catalog and research databases. If you need further assistance, please ask a librarian.

Books

Agents of Chaos: Earthquakes, Volcanoes, and Other Natural Disasters by Stephen L. Harris. Missoula, MT: Mountain Press, 1990. ISBN: 0878422439.
Focus on earthquakes and volcanoes in the Western United States.

At Risk: Natural Hazards, People's Vulnerability, and Disasters by Piers M. Blaikie. New York: Routledge, 1994. ISBN: 0415084776.
Focuses on what makes people vulnerable and looks at different social groups that suffer more in extreme events.

Crucible of Hazard edited by James K. Mitchell. Tokyo: UNU Press, 1999. ISBN: 9280809873.
Offers a compilation of maps that superimpose vulnerable populations with the physical hazard. Maps and lists are provided that identify municipalities in metro regions with high percentages of vulnerable people.

Devastation!: The World's Worst Natural Disasters by Lesley Newson. New York: Dorling Kindersley, 1998. ISBN: 0789435187.
Details from the world's most spectacular natural disasters. Includes 400 color photos.

Disasters by Design: A Reassessment of Natural Hazards in the United States by Dennis S. Mileti. Washington, DC: John Henry Press, 1999. ISBN: 0309063604.
Summarizes the hazards research findings from the previous two decades, synthesizes what has been learned, and outlines a proposed shift in direction in research and policy for natural and related technological hazards in the United States.

Encyclopedia of Earthquakes and Volcanoes by David Ritchie and Alexander E. Gates, PhD. New York: Facts On File, 2001. ISBN: 0816043728.
Science of seismology and volcanology through facts about the disasters themselves, historical and eyewitness accounts, and how they affect the political landscape.

Encyclopedia of Hurricanes, Typhoons, and Cyclones by David Longshore. New York: Facts On File, 1998. ISBN: 0816033986.
Covers all major aspects of tropical cyclone activity. More than 200 extensively cross-referenced A-to-Z entries detail cyclonic storms in meteorology, history, and culture.

The Environment as Hazard, 2nd ed. by Ian Burton, Robert William Kates, and Gilbert F. White. New York: Oxford University Press, 1993. ISBN: 019502222X.
Examines research and policy issues relating to natural hazards.

Natural Disasters, 4th ed. by Patrick L. Abbott. Boston: McGraw-Hill, 2003. ISBN: 0072921986.
Explores natural disasters such as volcanoes, earthquakes, flooding, and other hazards, how these phenomena develop, and their impact on the environment.

Natural Disasters by David E. Alexander. New York: UCL Press and Chapman & Hall, 1993. ISBN: 0412047411.
Overview of the physical, technological, and social components of natural disaster.

Web Sites

DisasterHelp
https://disasterhelp.gov/portal/jhtml/index.jhtml
Provides information and services relating to preparedness, response, recovery, and mitigation. Includes disaster news, anti-terror, disease, and natural disasters.

Federal Emergency Management Agency, Homeland Security
www.fema.gov
Leads the effort to prepare the nation for all hazards and manage federal response and recovery efforts following any national incident.

Forces of Nature
http://library.thinkquest.org/C003603/
Includes information regarding earth science, geology, fifteen common natural disasters, unusual phenomena, their impact, effects, and causes.

National Hazards Center
www.colorado.edu/hazards/
Collects and shares research and experience related to preparedness for, response to, recovery from, and mitigation of disasters.

Natural Disaster Reference Database
http://ndrd.gsfc.nasa.gov
A bibliographic database on research, programs, and results which relate to the use of satellite remote sensing for disaster mitigation. The NDRD was compiled and abstracted from articles published from 1981 though January 2000.

Natural Hazard Statistics
www.nws.noaa.gov/om/hazstats.shtml
Provides statistical information on fatalities, injuries and damages caused by weather related hazards.

Weather Disasters from InfoPlease
www.infoplease.com/weather.html?mc=1562837770
Provides details on some of the most deadliest weather disasters known.

Subject Headings

- avalanches
- disaster relief
- earthquakes
- emergency management
- fires
- floods
- hazardous geographic environments
- natural disasters
- typhoons
- volcanoes

Obituaries: A Research Guide

Obituaries provide an important tool to researchers by offering biographical information on individuals, which may not be readily available elsewhere. In many cases, the obituary supplies the only biographical data extant on a person. Major newspapers like *The New York Times* print obituaries of notable people and death notices for others in the local community. Indexes to these newspapers, as well as specialized indexes to obituaries in these and other sources, provide access to this valuable biographical information. In response to renewed interest in genealogy, various compilers have published indexes to obituaries that have appeared in American newspapers of modest circulation; obviously, these indexes are limited in terms of their geographical and chronological parameters. The terms and phrases listed in the subject headings below can be used to search for more materials in the library's catalog and research databases. If you need further assistance, please ask a librarian.

Books and Indexes

52 McGs: The Best Obituaries from Legendary New York Times Writer Robert McG. Thomas, Jr. edited by Robert McG. Thomas and Chris Calhoun. New York: Charles Scribner's Sons, 2001. ISBN: 0743215621.
The late *New York Times* reporter Robert McG. Thomas Jr. (1939–2000) developed a loyal following for quirky, witty obituaries.

Final Placement: A Guide to the Deaths, Funerals, and Burials of Notable Americans. Robert B. Dickerson. Algonac, MI: Reference, 1982. ISBN: 0917256182.
Information about the deaths, funerals, and burials of some famous Americans.

The Last Word: The New York Times Book Of Obituaries And Farewells: A Celebration Of Unusual Lives by Marvin Siegel. New York: Morrow, 1997. ISBN: 0688150152.
A compilation of 100 of the most colorful, entertaining, and touching obits that have appeared in *The New York Times.*

The New York Times Obituaries Index: 1858–1978. 2 vols. New York: New York Times, 1970–1980. ISBN: 0667005986.
Refers researchers to the appropriate issue of *The New York Times* for the desired obituary. After 1978, check the heading "Deaths" (which is then subdivided by individual names) in the annual cumulations of *The New York Times Index.*

Obituaries: A Guide to Sources by Betty Jarboe. Boston: G. K. Hall, 1989. ISBN: 0816104832.
Published obituaries and cemetery records arranged by state. Also indexed by names, titles, and subjects.

Obituaries in American Culture by Janice Hume. Jackson: University of Mississippi Press, 2000. ISBN: 1578062411.
A look at obituaries in American culture over time.

Obituaries on File by Felice D. Levy. 2 vols. New York: Facts On File, 1979. ISBN: 0871963728.
Compilation of obituaries appearing in *Facts On File* from 1940 through 1978. Arranged by name of the deceased, subject (wherein individuals are listed by vocation, nationality, and corporate affiliation) and chronology (in which the deceased are listed by date of death).

Permanent Addresses: A Guide to the Resting Places of Famous Americans by Jean S. Arbeiter and Linda D. Cirino. New York: Evans, 1983. ISBN: 0871314029.
Listings of where famous people are buried.

Project Remember: A National Index of Gravesites of Notable Americans by Arthur S. Koykka. Algonac: Reference, 1986. ISBN: 0917256220.
Arranged primarily by occupation with an index listing individual names. There are significant secondary arrangements which include mass burials, burials outside the United States, burials of uncertain disposition (e.g., those lost at sea), and a listing of those who chose to have no memorial e.g., those who directed that their ashes be scattered over land or sea).

Selected Internet Sites

Cemetery Junction
www.daddezio.com/cemetery/
Directory of cemeteries within the United States, Canada and Australia.

Family Search from The Church of Jesus Christ of Latter-Day Saints
www.familysearch.org/Eng/Search/frameset_search.asp
Search for death and burial records worldwide.

National Archives and Records Administration (NARA)
www.nara.gov
Online access to a selection of nearly 50 million historic electronic records created by more than 20 federal agencies.

Obituary Daily Times
www.rootsweb.com/~obituary/
Citations to published obituaries in newspapers in the United States and Canada.

Social Security Death Index
www.ancestry.com/search/rectype/vital/ssdi/main.htm
Generated from the U.S. Social Security Administrations Death Master File. It contains the records of deceased persons who possessed Social Security numbers and whose death had been reported to the SSA.

Vital Records Information-State Index
http://vitalrec.com/index.html
Contains information about where to obtain vital records (such as birth, death and marriage certificates, and divorce decrees) from each state, territory, and county of the United States.

Subject Headings

- cemeteries
- death
- death notices
- sepulchral monuments

On This Day in History: A Research Guide

Whether looking up your birth date or other days in history, it is interesting to find out what events took place locally, nationally, and around the globe. Equally as fascinating are the prices that people paid for cars and houses, as well as products advertised. The resources listed below will help researcher find both events and facts about days passed. This guide introduces the researcher to some of the basic informational sources on the topic. The terms and phrases listed in the subject headings below can be used to search for more materials in the library's catalog and research databases. If you need further assistance, please ask a librarian.

Books Day-by-Day

American Book of Days, 4th ed. by Stephen G. Christianson. New York: H. W. Wilson, 2000. ISBN: 0824209540.
Essays explore significant events for each day of the year including military, scientific, ethnic, and cultural events. An appendix features historical documents.

The Big Book of Dates: A Chronology of the Most Important People, Events, and Achievements of All Time by Laurie E. Rozakis. New York: McGraw-Hill, 2001. ISBN: 0071361022.
Chronological guide to the most important people, events, ideas, inventions, art, and achievements including fascinating sidebars about home, fashion, and fun, and excerpts from primary documents.

Chase's Calendar of Events. New York: McGraw-Hill, 2004. ISBN: 0071424059.
A calendar of over 12,000 listings with special days, weeks, and months as well as holidays, historical anniversaries, and fairs and festivals.

On This Date: A Day-by-Day Listing of Holidays, Birthday and Historic Events, and Special Days, Weeks and Months edited by Sandra Whiteley, et al. Chicago: Contemporary Books, 2003. ISBN: 0071398279.
Comprised of select entries from the Chase's Calendar of Events, this handy, fact-filled reference is packed with more than 2,000 events, milestones and trivia.

The Oxford Book of Days by Bonnie J. Blackburn and Leofranc Holford-Strevens. New York: Oxford University Press, 2000. ISBN: 0198662602.
Includes historical facts, legend, lore, and literature listed for each season, month, and day of the calendar.

Chronologies

America in the 20th Century, 2nd ed. 13 vols. Tarrytown, NY: Marshall Cavendish, 2003. ISBN: 0761473645.
Organized by decade and features photos and art, vols. 1–10 consider important people and trends in American politics, social policy and civil rights, foreign policy, popular culture and recreation, economy and trade, literature and arts, and the environment. A selection of over one hundred primary sources are contained in vols. 11 and 12. Grades 7 and up.

America's Century: Year by Year from 1900 to 2000 by Tom Anderson and Clifton Daniel. New York: Dorling Kindersley, 2000. ISBN: 0789453398.
Events of the 20th century are recounted year-by-year in newspaper-style reports. Also includes over 1,300 full-color photographs and drawings.

American Chronicle: Seven Decades in American Life, 1920–1989 by Lois G. Gordon and Alan Gordon. New York: Crown, 1990. ISBN: 0517575752.
Highlights of each decade are reviewed including vital economic, social, and consumer statistics.

American Decades by Vincent Tompkins, et al. 10 vols. Detroit: Gale Research, 1994–2001. ISBN: 0810357224. Vol. 1: 1900–1909; vol. 2: 1910–1919; vol. 3: 1920–1929; vol. 4: 1930–1939; vol. 5: 1940–1949; vol. 6: 1950–1959; vol. 7: 1960–1969; vol. 8: 1970–1979; vol. 9: 1980–1989; vol. 10: 1990–1999.
Documents and analyzes periods of contemporary American social history. Presents overview of time period and chronology covering the entire decade including an explanatory background overview; subject-specific time line; and alphabetically arranged entries discussing the people, events and ideas important to that subject during the period.

The People's Chronology: A Year-by-Year Record of Human Events from Prehistory to the Present by James Trager. New York: Henry Holt, 1994. ISBN: 0805031340.
Comprehensive reference detailing everyday social and political life.

The Timetables of History: A Horizontal Linkage of People and Events, 3rd ed. by Bernard Grun and Werner Stein. New York: Simon & Schuster, 1991. ISBN: 0671749196.
Lists more than 30,000 events in an overview of 7,000 years of civilization highlighting significant moments in history, politics, philosophy, religion, art, science, and technology.

Web Sites

Anyday: Today in History
www.scopesys.com/anyday/
Provides brief listing of historic events that occurred on a selected days.

dMarie Time Capsule
http://dmarie.com/timecap/
Enter a date and create a time capsule that highlights events, famous people born on that date, popular music, movies, books, TV shows and toys during that time.

The New York Times: On This Day
www.nytimes.com/learning/general/onthisday/index.html
Provides brief information on events and photographs of famous people.

This Day in History
www.historychannel.com/today/
Select a date and retrieve general interest or any of the following categories: automotive, Civil War, cold war, crime, entertainment, literary, Old West, Vietnam War, Wall Street, World War II. Provides more detailed information.

Today in History from Library of Congress
http://lcweb2.loc.gov/ammem/today/today.html
Provides in-depth information on national historical events.

Subject Headings

- calendar
- calendars—history
- chronology, historical
- United States—civilization—date
- United States—history—chronology

Patents, Trademark, and Copyright: A Research Guide

People research the existence of patents in hopes of inventing something unique or improving on a product that has already been invented. They also use patent research to assist them in completing their patent applications to the United States Patent and Trademark Office. Trademark searches uncover the registered use of a particular name to identify a product, service, or business. Whether starting a new business, creating a new product, or writing a book, the laws on intellectual property remain a maze for most of us. The materials and resources listed below will assist researchers with a starting off point to find further materials.

Books

The Copyright Handbook: How to Protect & Use Written Words, 7th ed. by Stephen Fishman. Berkeley, CA: Nolo Press, 2003. ISBN: 0873379748.
Provides you with all the information and forms that you'll need to protect all types of creative expression under U.S. and international copyright law.

How to Make Patent Drawings Yourself: Prepare Formal Drawings Required by the U.S. Patent Office, 3rd ed. by Jack Lo and David Pressman. Berkeley, CA: Nolo Press, 2002. ISBN: 0873377885.
Explains how to create patent drawings that comply with the rules of the U.S. Patent Office.

Patent, Copyright & Trademark: An Intellectual Property Desk Reference, 6th ed. by Stephen Elias and Richard Stim. Berkeley, CA: Nolo Press, 2003. ISBN: 0873372360.
Provides overview of intellectual property law in practical terms.

Patent It Yourself, 10th ed. by David Pressman. Berkeley, CA: Nolo Press, 2004. ISBN: 1413300251.
Guides readers through the entire patent process, from conducting a patent search to filing a successful application.

Trademark: Legal Care for Your Business & Product Name, 6th ed. by Stephen Elias. Berkeley, CA: Nolo Press, 2003. ISBN: 0873379454.
Guide on choosing and protecting your business name whether it's in print or on the Web.

Associations

Intellectual Property Owners Association (IPO)
www.ipo.org
Established in 1972, IPO is a trade association for owners of patents, trademarks, copyrights and trade secrets. IPO is the only association in the United States that serves all intellectual property owners in all industries and all fields of technology.

National Association of Patent Practitioners (NAPP)
www.napp.org
A nonprofit organization dedicated to supporting patent practitioners and those working in the field of patent law in matters relating to patent law, its practice, and technological advances.

National Congress of Inventor Organizations (NCIO)
http://inventionconvention.com/ncio/
Oldest nonprofit umbrella organization in the United States dedicated to encouraging the spirit of innovation and creativity. NCIO offers education, training, and support to inventor organizations with the mission of creating standards of ethics and professionalism within the invention community and growing the invention industry.

United Inventors Association
www.uiausa.org
Not for profit corporation formed in 1990 for educational purposes. Its mission is to provide leadership, support, and services to inventor support groups and independent inventors.

Web Sites

Canadian Patents Database
http://patents1.ic.gc.ca/intro-e.html
This database lets you access over 75 years of patent descriptions and images. You can search, retrieve and study more than 1,500,000 patent documents.

European Patent Office
www.european-patent-office.org/online/
Links to member states and patent databases.

Frequently Asked Questions About Patents
www.uspto.gov/web/offices/pac/doc/general/faq.htm
Frequently asked questions and answers about patents.

Frequently Asked Questions About Trademarks
www.uspto.gov/web/offices/tac/tmfaq.htm
Frequently asked questions and answers about trademarks.

Patent and Trademark Depository Library Program (PTDLP)
www.uspto.gov/web/offices/ac/ido/ptdl/ptdlib_1.html
Find a Patent and Trademark Depository Library (PTDL) near you. These libraries receive and house copies of U.S. patents and patent and trademark materials. They have staff that can guide you through the patent search process.

Patent Cafe
www.patentcafe.com
Covers many practical aspects of intellectual property including patent searching, invention groups, government agencies, and famous women, and black inventors. There are chat groups, magazines, and book sources.

United States Patent Office
www.uspto.gov
Official Web site of the U.S. Patent Office. Search for existing patents, instructional information, and online filing.

Subject Headings

- copyright
- patent laws and legislation
- trademarks—law and legislation

Personal Finance: A Research Guide

Looking for tips on how to make your money grow or on how to spend it more wisely? The resources listed below will assist researchers in making wise decisions like where to bank their money and where to invest it. This guide introduces researchers to some of the basic informational sources on the topic. The terms and phrases listed in the subject headings below can be used to search for more materials in the library's catalog and research databases. If you need further assistance, please ask a librarian.

Banking and Bill Paying

Bank Rate
www.bankrate.com
List of best interest rates and a wide array of financial calculators. Separate forums for the following topics: tax, small business, investment and advice.

Checkfree
www.checkfree.com
Use this service to receive and pay bills electronically. Requires paid subscription.

FDIC
www.fdic.gov/bank/index.html
Searchable databases allow users to find institutions and their branches in order to retrieve contact and financial status.

National Credit Union Administration
www.ncua.gov
Provides directory of domestic and international credit unions with links to each institution.

Paytrust
www.paytrust.com
Allows you to receive and review all your bills online, decide who to pay, how much and when and from which bank account. Requires paid subscription.

Insurance

Best's Ratings Online
www.ambest.com/ratings/access.html
Includes contact information and brief financials on your insurance company. You can also find a table that defines the Best rating and what it actually means.

InsWeb
www.insweb.com
Allows you to compare insurance quotes from leading insurance companies to find the best rates available for vehicles, home, life, medical and pet insurance.

Brokerage Services

The following commercial investment services offer a variety of products and services including the ability to buy and sell stock as well as basic investment information. They allow you to view your portfolio, research industries, markets, and individual stocks.

Ameritrade: *www.ameritrade.com*

Datek: *www.datek.com*

DLJ Direct: *www.dljdirect.com*

E*trade: *www.etrade.com*

News and Magazines

CBS Marketwatch
http://cbs.marketwatch.com
Enormous amount of information best navigated with a ticker symbol in hand. The site provides, as it says, "the story behind the numbers," with headlines and news on companies, industries, markets, and investment topics.

CNN FN The Financial Network
http://cnnfn.cnn.com
Financial news including sections on technology, companies, CEOs, the economy, and industry.

Forbes
www.forbes.com
Provides articles from its current and archived magazine. Forbes famous lists are conveniently located on one page. Find the "World's Richest People," "Forbes 500," "500 Largest Private Companies," "Forbes Platinum 400," and "Forbes 200 Best Small Companies."

Web Sites

BusinessWeek Online, Investing
www.businessweek.com/investor/index.html
A wealth of information on markets, stocks, funds, economy, IPOs, industries, and current news in the investment world.

Fool.com
www.fool.com
Includes stock research, quotes, data and discussion boards. Offers a customizable portfolio tracker and weekly stock screen to generate new investment opportunities.

MSN MoneyCentral
http://moneycentral.msn.com
Offers a wide array of information for the Investor including banking and bill paying, retirement, wills, taxes, insurance, saving, spending, family, college, mortgages, and loans.

Smartmoney.com
www.smartmoney.com
Offers a comprehensive site for beginning and seasoned investors with many interactive tools.

Stock Screener
www.quicken.com/investments/stocks/search/
Use Stock Search to find stocks that best meet your investing goals. Automated searches include preset criteria that match popular investing strategies or use the full search to create your unique stock screener by combining certain variables.

Yahoo! Finance
http://finance.yahoo.com
Provides a directory of investing and personal finance Web sites, including top business news for the day and links to other Yahoo! International market sites. Also provides a quick market summary and brief synopsis of the day's trading activities.

Subject Headings

- corporations—United States
- finance—periodicals
- investments—periodicals
- mutual funds
- securities

Psychology: A Research Guide

A relatively new field that came into its own during the 19th century, psychology has its roots in philosophy, physiology, and psychiatry and includes the study of human and animal behavior. Psychology can be categorized by the following sub fields: cognitive, abnormal, social, and developmental psychology. This document serves as a starting point and identifies some of the more general resources available for students and researchers. The terms and phrases listed in the subject headings below can be used to search for more materials in the library's catalog and research databases. If you need further assistance, please ask a librarian.

Encyclopedias and Dictionaries

Biographical Dictionary of Psychology edited by Noel Sheehy, Antony J. Chapman, and Wendy A. Conroy. New York: Routledge Reference, 2002. ISBN: 0415099978.
Includes basic information on 500 individuals who have had significance influence in the field of psychology. Lists birth, nationality, education, awards, and publications.

Diagnostic and Statistical Manual of Mental Disorders: DSM-IV, 4th ed. Washington, DC: American Psychiatric Association, 1994. ISBN: 0890420610.
Provides a classification of mental disorders (e.g., anxiety disorders, eating disorders, paranoid disorders, etc.) and attempts to describe in detail their symptoms, manifestations, and etiology when known.

The Dictionary of Psychology by Raymond J. Corsini. Philadelphia: Brunner/Mazel, 1999. ISBN: 158391028X.
Brief definitions of over 10,000 psychological terms.

Encyclopedia of Human Behavior edited by V.S. Ramachandran. 4 vols. San Diego: Academic Press, 1994. ISBN: 0122269209.
250 signed articles arranged alphabetically includes glossary, introduction on topic, and bibliographies. Topics range from adolescence to brainwashing and totalitarian influence.

The Encyclopedia of Psychology edited by Alan E. Kazdin. 8 vols. Oxford: Oxford University Press, 2000. ISBN: 1557986509.
Covers all areas of psychological theory, research, and practice including concepts, methods, theories, findings, major figures, schools of thought, and emerging areas of interest. Produced by the American Psychological Association and Oxford University Press.

The Gale Encyclopedia of Psychology, 2nd ed. by Bonnie B. Strickland. Detroit: Gale Group, 2001. ISBN: 0787647861.
Includes 500 entries on topics covering key concepts in psychology. Provides definition of term preceding each entry. Illustrations accompany entries where appropriate.

The Mental Measurements Yearbook edited by Oscar Krisen Buros. Highland Park, NJ: Mental Measurements Yearbook, 1940– . ISSN: 00766461.
Designed to assist users in the fields of education, psychology and industry to make intelligent use of standardized tests. It provides critical descriptions of these tests as well as extensive bibliographic references to books and articles about them.

The Oxford Companion to the Mind edited by Richard L. Gregory and Oliver L. Zangwill. New York: Oxford University Press, 1987. ISBN: 019866124X.
Signed articles ranging in length from extensive (Freud, intelligence) to brief (halo effect, shock). Some entries include bibliographies.

Survey of Social Science (Psychology Series) edited by Frank Northen Magill and Jaclyn Rodriquez. 6 vols. Pasadena, CA: Salem Press, 1993. ISBN: 0893567329.
Provides general reader with insight into psychology topics with 410 articles including topic significance, key terms and definitions, overview, application, context an bibliography. Topics are more broad than specific.

Journals

American Psychologist
APA Monitor
Contemporary Psychology
Psychological Assessment
Psychological Bulletin
Psychological Methods
Psychological Review

See also:
Journals in Psychology: A Resource Listing for Authors. American Psychological Association, 5th ed. Washington, DC: American Psychological Association, 1997. ISBN: 1557984387.

Web Sites

PsychCrawler
www.psychcrawler.com
A search engine providing an index to psychology-related Web sites.

Psychology Online Resource Central
www.psych-central.com
Provide links to Web sites on categories like conventions, discussion groups, online journals, and organizations on the Web.

PSYCHWEB
www.psychwww.com
Web site dedicated to psychology. Written for students and their professors.

Sigmund Freud: Conflict and Culture
http://lcweb.loc.gov/exhibits/freud/
Library of Congress exhibition on Freud that includes selected film and television clips and materials from newspapers, magazines, and comic books.

Organizations

American Psychological Association
www.apa.org
Works to advance psychology as a science, a profession, and as a means of promoting human welfare. Scientific and professional society of psychologists.

American Psychological Society
www.psychologicalscience.org
Works for the advancement of the discipline of psychology and the promotion of human welfare through research and application.

State and Provincial Psychological Associations
www.apa.org/practice/refer.html
Lists contact information to associations in the United States and Canada.

Subject Headings

- child psychology
- cognitive psychology
- human behavior
- psychologists—biography—dictionaries
- psychology
- psychology—history
- psychology, applied

Public Administration and Policy: A Research Guide

Public administration is concerned with the formulation and implementation of government policies and programs. Its concepts and methods come from political science, sociology, psychology, economics, and organizational studies. Public policy and public management are two aspects of public administration. Public policy focuses on the formulation and analysis of government policies and public management examines the implementation and evaluation of government programs. This guide introduces researchers to some of the basic informational sources on the topic. The terms and phrases listed in the subject headings below can be used to search for more materials in the library's catalog and research databases. If you need further assistance, please ask a librarian.

Reference Books

Encyclopedia of Policy Studies edited by Stuart S. Nagel. New York: Marcel Dekker, 1994. ISBN: 0824791428.
Provides general approaches to policy studies and policy problems.

Encyclopedia of Public Administration and Public Policy edited by Jack Rabin. 2 vols. New York: Marcel Dekker, 2003. ISBN: 0816047995.
Provides short essays on various topics related to public administration or public policy.

Facts On File International Encyclopedia of Public Policy and Administration edited by Jay M. Shafritz. 4 vols. Boulder: Westview Press, 1998. ISBN: 0813399734.
Encyclopedia of concepts and terms arranged alphabetically. Brief bibliographies often accompany entries.

Handbook of Comparative and Development Public Administration, 2nd ed. by Ali Farazmand. New York: Marcel Dekker, 2001. ISBN: 0824704363.
Examination and analysis of public administration.

Handbook of Public Administration, 2nd ed. edited James L. Perry. San Francisco: Jossey-Bass, 1996. ISBN: 0787901946.
Overview of the public administration field including relationships with other government branches, policies and programs, budgets and fiscal administration, human resources, operations and services, and the profession.

Handbook of Public Administration, 2nd ed. edited by Jack Rabin, W. Bartley Hildreth, and Gerald Miller. New York: Marcel Dekker, 1998. ISBN: 0585138583.
Examines all major areas in public administration including public budgeting, financial management, decision making, public law and regulation, and political economy

Handbook of Public Law and Administration edited by Phillip J. Cooper and Chester A. Newland. San Francisco: Jossey-Bass, 1997. ISBN: 0787909300.
Provides an overview of all aspects of public law and how it affects the public administrator's job and responsibilities.

The Public Administration Dictionary, 2nd ed. edited by Ralph C. Chandler and Jack C. Plano. Santa Barbara, CA: ABC-CLIO, 1988. ISBN: 0874364981.
Includes concepts and terms that are fundamental to an understanding of the field of public administration. Defined terms are arranged in seven subject categories with indexes both for those terms and others included in the text of definitions.

INDEXES AND ABSTRACTS

Public Affairs Information Service Bulletin (PAIS). New York: Public Affairs Information Service, 1968– . Monthly. ISSN: 10514015.
Indexes resources in public affairs and policy for use by the government, the business/financial community, researchers and students. Emphasis is on factual and statistical information. 1400 periodicals, as well as books, reports, and federal and state documents are arranged by subject and indexed by author.

Sage Public Administration Abstracts. Newbury Park, CA: Sage, 1974– . Quarterly. ISSN: 00946958.
Contains abstracts from important recent literature in public administration. Included are abstracts of books, articles, pamphlets, government publications, speeches, and legislative research studies.

ASSOCIATIONS

American Society for Public Administration
www.aspanet.org
Professional association for public administrators for government and nonprofit organizations.

Institute of Public Administration
www.theipa.org
Organization concerned with building capacity for effective government through programs of research, technical assistance, and training in three areas: public sector governance and management, public finance and fiscal reform, and sustainable urban development.

National Academy of Public Administration
www.napawash.org
Information about the organization, news, projects, and publications.

SUBJECT HEADINGS

- policy making
- policy sciences
- public administration
- public administration—cross-cultural studies
- public administration—handbooks, manuals, etc.
- public administration—United States
- public policy

Quotations: A Research Guide

Dictionaries of quotations fulfill many needs: identifying a given quotation, verifying an author, providing suitable quotations for a subject area, and supplying selections from the writings of prominent persons. The library provides a wide assortment of general and specialized quotation dictionaries, and these sources, when properly employed, provide the researcher with an inestimable tool in researching the statements of notable people. This guide introduces researchers to some of the basic informational sources on the topic. The terms and phrases listed in the subject headings below can be used to search for more materials in the library's catalog and research databases. If you need further assistance, please ask a librarian.

Dictionaries of Quotations

Familiar Quotations: A Collection of Passages, Phrases, and Proverbs Traced to Their Sources in Ancient and Modern Literature, 16th ed. by John Bartlett and Justin Kaplan. Boston: Little, Brown, 1992. ISBN: 0316082775.
A standard, comprehensive collection whose primary arrangement is by author (listed chronologically by birth date); indexes are by keyword and author. Earlier editions are useful for quotations omitted in the latest edition.

The Home Book of Proverbs, Maxims, and Familiar Phrases by Burton Egbert Stevenson. New York: Macmillan, 1948.
This comprehensive work is arranged by subject with cross-references to related headings; a keyword index is provided.

The Home Book of Quotations, Classical and Modern, 10th ed. by Burton Egbert Stevenson. New York: Dodd, Mead, 1967.
This comprehensive dictionary of quotations is primarily arranged by subject; author and keyword indexes are provided.

International Thesaurus of Quotations by Rhoda T. Tripp. New York: Crowell, 1970. ISBN: 0690445849.
Primarily arranged by subject, this work is also indexed by author, source, and keyword.

The New Beacon Book of Quotations by Women compiled by Rosalie Maggio. Boston: Beacon Press, 1998. ISBN: 0807067830.
16,000 quotes from 2,600 women—most of them found in no other collection.

A New Dictionary of Quotations on Historical Principles from Ancient and Modern Sources by Henry L. Mencken. New York: Knopf, 1942.
This comprehensive collection, notable for its inclusion of many lesser-known quotations, is arranged only by subject (with cross-references to related headings). No indexes.

The Oxford Dictionary of Modern Quotations by Tony Augarde. New York: Oxford University Press, 1991. ISBN: 019866141X.
Arranged by author, this dictionary supplies a keyword index.

The Quotable Woman, from Eve to 1799 (1985; ISBN: 0871963078) and *The Quotable Woman, 1800–1981* (1982; ISBN: 0871965801) compiled by Claudia B. Alexander. New York: Facts On File.
These companion volumes are arranged chronologically by birth date of author; indexes are by author and subject.

Quotations in Black by Anita King. Westwood, CT: Greenwood Press, 1981. ISBN: 0313221286.
Arranged chronologically by birth date of author. Indexed by author, subject, and keyword.

The Wisdom of the Novel: A Dictionary of Quotations by David Powell. New York: Garland, 1985. ISBN: 0824090179.

Arranged by subject, this dictionary draws its quotations from English and American authors published between 1470 and 1900; author, novel, and keyword indexes are provided.

Web Sites

Bartlett's Familiar Quotations
www.bartleby.com/100/
Search and browse 11,000 quotations from the 10th edition of *Bartlett's* published in 1919.

IMDb Movie/TV Quotes Browser
www.imdb.com/Sections/Quotes/
Movie and TV quotes from the Internet Movie Database.

Quoteland.com
www.quoteland.com/
Literary, humorous, and random quotations, plus quotations by topic.

Simpson's Contemporary Quotations
www.bartleby.com/63/
10,000 notable quotations from 4,000 sources covering the years 1950 to 1988. Browse subject and author indexes to find quotations with full citations.

Subject Headings

Find books about quotations by using the search term "quotations" and names of specific individuals (Albert Schweitzer quotations), language and nationality (Quotations—American, Quotations—Russian) or by subjects.

- law—quotations
- presidents—United States—quotations
- religion—quotations
- sports—quotations
- Wall Street—quotations
- women—quotations

Religion: A Research Guide

Students of religion will find a starting point for their research by using the resources listed below. Research materials can include primary texts of religious writings and beliefs including holy books (Torah and the Talmud; Qur'an, Bible, Guru Granth Sahib), teachings, symbols, and contact information. Additionally, researchers can find information on comparing world religions and government tolerance of religions. The terms and phrases listed in the subject headings below can be used to search for more materials in the library's catalog and research databases. If you need further assistance, please ask a librarian.

Encyclopedias

Contemporary American Religion edited by Wade Clark Roof. 2 vols. New York: Macmillan Library Reference, 2000. ISBN: 0028649281.
Over 500 well-written entries by 250 contributors cover the major religions such as Catholicism and Judaism, along with cults and phenomena such as Heaven's Gate and Cyber Religion.

Eerdmans Dictionary of the Bible edited by David Noel Freedman. Grand Rapids, MI: Eerdmans, 2000. ISBN: 0802824005.
Covers 5,000 articles including definitions, personal names, and their derivation, places, and concepts from the Bible from 600 leading biblical scholars.

Encyclopedia of Christianity edited by Erwin Fahlbusch. 3 vols. Grand Rapids, MI: Eerdmans, 1999. ISBN: 0802824137. Vol. 1. A–D—vol. 3. J–O. (3 of 5 vols.)
This multivolume set is a translation of the third revised edition of the German work, *Evangelisches Kirchenlexicon*, with articles added and expanded for English-speaking readers. This encyclopedia describes Christianity through its 2,000-year history within a global context. Articles have been added and expanded for English-speaking readers. Volumes four and five of the series are forthcoming.

New Catholic Encyclopedia, 2nd ed. Catholic University of America. 15 vols. Detroit: Washington, DC: Thomson/Gale; Catholic University of America, 2003. ISBN: 0787640042.
Covers persons and subjects related to Catholicism and the humanities. Covers religious aspects of contemporary issues such as abortion, divorce, cloning, and reproductive technology.

The New Encyclopedia of Judaism by Geoffrey Wigoder, Fred Skolnik, and Shmuel Himelstein. New York: New York University Press, 2002. ISBN: 0814793886.
Encyclopedia that presents every aspect of the Jewish religion and represents current thinking among scholars in the Reform, Conservative, and Orthodox movements.

New Historical Atlas of Religion in America. (Rev. ed. of: *Historical Atlas of Religion in America* by Edwin Scott Gaustad. 1976.) by Edwin Scott Gaustad and Philip L. Barlow. Oxford: Oxford University Press, 2001. ISBN: 019509168X.
Includes 260 colorful, detailed maps and 200 other graphics provide the histories, migration, developments, and growths of religious communities in the United States.

Religions of the World: A Comprehensive Encyclopedia of Beliefs and Practices by J. Gordon Melton and Martin Baumann. 4 vols. Santa Barbara, CA: ABC-CLIO, 2002. ISBN: 1576072231.
Religious history and key religious communities in all of the more than 240 recognized nations and territories of the world.

Web Sites

American Religion Data Archive
www.thearda.com
The ARDA collection includes data on churches and church membership, religious professionals, and religious groups (individuals, congregations, and denominations).

Guide to the Religions of the World
www.bbc.co.uk/worldservice/people/features/world_religions/
Covers major religions of the world: Buddhism, Hinduism, Christianity, Islam, Judaism, Sikhism; their history and current status.

International Religious Freedom, U.S. Department of State
www.state.gov/g/drl/rls/irf/2003/index.htm
Annual report on International Religious Freedom supplementing the most recent Human Rights Reports by providing additional detailed information with respect to matters involving international religious freedom. This annual report includes individual country chapters on the status of religious freedom worldwide.

Virtual Religion Index
http://religion.rutgers.edu/vri/
Provides links and descriptions to hundreds of religion Web sites.

Subject Headings

Researchers can search by the name of the religion and add terms like dictionary or encyclopedia.

- Buddhism
- Catholicism
- Christianity
- Hinduism
- Islam
- Judaism
- religions

Research Tools: A Research Guide

The tools listed below will help answer many of the questions that can be answered by a quick reference to an encyclopedia, dictionary, almanac, or by a telephone call for information. Resources listed below help you find area codes, zip codes, translations, time zones, and currency conversions. This guide introduces researchers to some of these basic informational sources. If you need further assistance, please ask a librarian.

Area Codes

555-1212.com
www.555-1212.com
Looking to match an area code with the name of a city? Look no further.

Maps and Directions

Mapquest
www.mapquest.com
Easy access to driving directions from one point to another, customized maps, road trip planner, and local traffic reports.

MSN Maps and Directions
www.mapblast.com
Offers directions and maps internationally. Search by address or place name.

Yahoo! Maps
http://maps.yahoo.com
Simple searches for maps and directions.

Calculations and Conversions

ConvertIt: Conversion and Calculation Center
www.convertit.com
Provides measurement conversions, currency converter, world time clock, and links to a variety of calculators.

Almanacs, Dictionaries, and Encyclopedias

Bartleby.com—Great Books Online
www.bartleby.com/reference/
Search the following reference books simultaneously: *Columbia Encyclopedia*, *World Factbook*, *American Heritage Dictionary of the English Language*, *Roget's II: The New Thesaurus,* and many more classics.

Britannica.com
www.britannica.com
Searchable encyclopedia entries, expert reviews of the Web's best sites, timely articles from leading magazines, and related books.

Dictionary.com
www.dictionary.com
Gives pronunciation, part of speech, definition, and history of the word from a variety of traditional print dictionaries. Includes thesaurus.

Free Translation
www.freetranslation.com
Translates English to Spanish, French, German, Italian, Norwegian, and Portuguese. Just type your text and press "Translate!"

Information Please
www.infoplease.com
Facts, dates, names and information on the following categories: world, history, government, biographies, sports, entertainment, business, and weather.

Old Farmer's Almanac
www.almanac.com
Gives weather updates for your region for this month and the next. Find the rising and setting sun times, gives best fishing days, and records information on gardening, astronomy, weather, food, and more.

Wordsmyth
www.wordsmyth.net
Combines the functions of a dictionary and a thesaurus by retrieving pronunciation, definition, synonyms, and grammatical information.

World English Dictionary
www.dictionary.msn.com
Provides definition and audio file of the spoken word to help you with pronunciation.

Calendars, Times and Dates

The Official U.S. Time
www.time.gov
Gives the official time for all zones in the United States.

World Time Server
www.worldtimeserver.com
Provides current local times by country or state in one click. Makes real time adjustments for Daylight Saving Time.

Thesaurus

Roget's II: The New Thesaurus
www.bartleby.com/62/
Contains 35,000 synonyms in an easy-to-use format. Part of the Bartleby.com collection.

Thesaurus.com
www.thesaurus.com
Search or browse by word or category to retrieve similar words. Also includes antonym finder.

Zip Codes

Canada Post
www.canadapost.ca
Look up postal codes for any address in Canada and find postal rates. Also look up address ranges for given postal codes.

PostInfo
www.postinfo.net/links/
Need help finding a postal code in another country? Visit this site. Lists official Web sites of the postal authorities around the world. Includes rate calculators.

U.S. Postal Service
www.usps.com
Find zip codes or find the city or state associated with a zip code.

Saints: A Research Guide

All of the world's main religions recognize saints, holy persons who becomes religious heroes by virtuous actions and deeds. Depending on the religion, saints are revered in different ways. Patron saints represent such things as animals, occupations, diseases, and days of the year. For example, did you know that Brendan the Navigator is the patron saint of whales or that Catherine of Alexandria is the patron saint of libraries? Use the references below to find out who your patron saint is! This guide introduces researchers to some of the basic informational sources on the topic. The terms and phrases listed in the subject headings below can be used to search for more materials in the library's catalog and research databases. If you need further assistance, please ask a librarian.

Reference Books

Butler's Lives of the Saints by Alban Butler, et al. Collegeville, MN: Liturgical Press, c1995–2000. Bibliographic information varies. ISBN: 0814623778
The 200-year-old series has undergone a thorough revision and rewriting in the most recent version and is now presented as a 12-volume set that corresponds to the months of the year—each volume contains entries on saints with feast days in that month.

Dictionary of Saints, 2nd ed. by John J. Delaney. Garden City, NY: Doubleday, 2003. ISBN: 0385135947.
Includes more than five thousand biographies of the saints, including martyrs, monks, mystics, virgins, doctors, and peasants. More in-depth listings for popular saints and thumbnail sketches for those less known. Includes the newly canonized saints, a complete listing of feast days, and an index of patron saints.

Lives of the Saints: from Mary and St. Francis of Assisi to John XXIII and Mother Teresa by Richard P. McBrien. San Francisco: HarperSanFrancisco, 2003, 2001. ISBN: 0060653418.
An A-to-Z guide on the lives of the saints. Emphasis on sanctity, historical and theological histories of the saints.

Oxford Dictionary of Saints, 5th ed. by David Hugh Farmer. New York: Oxford University Press, 2003. ISBN: 019860629X.
Includes accounts of the lives, cults, and artistic associations of over 1,400 saints. This edition includes a new appendix on pilgrimage sites in Europe. Includes only those saints venerated in the British Isles.

Books for Children

New Picture Book of Saints: Illustrated Lives of the Saints for Young and Old by Lawrence G. Lovasik. New York: Catholic Book, 1979. ISBN: 0899422357.
Illustrative stories of the lives of 106 popular saints. Aimed at elementary and junior high school students.

A Saint for Your Name: Saints for Boys (ISBN: 0879733306) and *A Saint for Your Name: Saints for Girls* (ISBN: 0879733314) by Albert J. Nevins and James McIlrath. Huntington, IN: Our Sunday Visitor, 1980.
Includes brief information on two hundred of the most popular saints' names for boys and girls.

Web Sites

Catholic Online Saints—Patron Saints
www.catholic.org/saints/
This site offers an index to saints arranged alphabetically or by nationality. You are given a brief biography of the saint as well as links to other biographical sites for many of these saints.

Hagiography database
www.doaks.org/hagio.html

Contains bio-bibliographical information about saints from the 8th–10th centuries. Divided into three sections: saints' lists, authors' lists, and a searchable section of citations.

Index of Saints
www.saintpatrickdc.org/ss/ss-index.htm

Use this index to locate by name the saints and beati included on this Web site. Also lists saints by their feast day.

Patron Saints Index
www.catholic-forum.com/saints/indexsnt.htm

A collage of information representing the biographies and stories of patron saints. Researchers can access information several different ways including by time line, calendar, patron saint name, and topic.

Subject Headings

- Catholic saints
- Christian saints—biography—dictionaries
- Christian saints—biography—juvenile literature
- patron saints
- saints
- saints—biography—dictionaries

Science Experiments: A Research Guide

Looking to score an "A" on your science project this year? Look no further than the books and Web sites listed below. Find great ideas and instructions on how to build the best science experiment ever! This guide introduces researchers to some of the basic informational sources on the topic. The terms and phrases listed in the subject headings below can be used to search for more materials in the library's catalog and research databases. If you need further assistance, please ask a librarian.

Books

Ideas for Science Projects by Robert Gardner. New York: F. Watts, 1986. ISBN: 0531102467.
Introduces the scientific method through instructions for observations and experiments in biology, physics, astronomy, botany, psychology, and chemistry.

The Complete Handbook of Science Fair Projects by Julianne Blair Bochinski and Judy J. Bochinski-DiBiase. Hoboken, NJ: J. Wiley, 2004. ISBN: 0471460435.
Discusses various aspects of science fair projects including advice on choosing a topic, doing research, developing experiments, organizing data results, and presenting a project to the judges.

Janice VanCleave's Guide to the Best Science Fair Projects by Janice Pratt VanCleave. New York: Wiley, 1997. ISBN: 0471148024.
Learn how to develop a topic from any idea, and find out the best ways to create, assemble, and present projects—including special tips on how to display them. Project ideas from astronomy and biology to chemistry, math, and engineering.

Science Experiments on File. 2 vols. New York: Facts On File, 2000. ISBN: 08160039984.
Provides substantial coverage of subjects in all the scientific disciplines studied in schools, from biology and chemistry to Earth science, physics, and more. Loose-leaf publication includes hundreds of illustrations. Volume 1 contains Earth science, weather, space, and biology; volume 2 contains chemistry and physics.

Science Experiments You Can Eat by Vicki Cobb and Peter J. Lippman. Philadelphia: Lippincott, 1972. ISBN: 0397314876.
Experiments with food demonstrate various scientific principles and produce an eatable result. Includes fruit drinks, grape jelly, muffins, chop suey, yogurt, and junket.

Science Projects and Activities by Helen J. Challand and Linda Hoffman Kimball. Chicago: Children's Press, 1985. ISBN: 0516005693.
Gives instructions for science projects and experiments involving an ant colony, a chicken brooder, plant grafting, microphotography, water purification, and other topics.

Super Sensational Science Fair Projects by Michael A. Dispezio and Derek Toye. New York: Sterling, 2003. ISBN: 0806944099.
Includes science fair projects for elementary and junior high school students.

For a list of more books see the science fair organizer located at *http://school.discovery.com/sciencefaircentral/scifairstudio/resources.html.*

Web Sites

Brainpop, Science
www.brainpop.com/science/seeall.weml
Suggest experiments and provides scientific information that can help support your project.

Bunsen Bob's Science Hunt
www.sciencehunt.com
Offers science projects and experiments for grades up to 8th and 9–12th. Includes ideas, instructions, and the best ways to display your projects.

California State Science Fair
www.usc.edu/CSSF/
This site highlights the final science fair of the academic year for students throughout the State of California in grades 6–12th. Includes information about previous science fairs.

IPL: Science Fair Project Resource Guide
www.ipl.org/div/kidspace/projectguide/
Explains the scientific methods, how to create a great science project and links to other Web sites.

Science Fair Central
http://school.discovery.com/sciencefaircentral/
Guide to science projects included project ideas and lists of Web sites and books to help you along the way.

Science Fair Project Ideas: Earthquakes
http://earthquake.usgs.gov/4kids/sciencefair.html
Ideas for science projects that are about earthquakes.

What Makes a Good Science Fair Project?
www.usc.edu/CSSF/Resources/Good_Project.html
From the California State Science Fair Judging Policy Advisory Committee.

SUBJECT HEADINGS

- science—exhibitions—handbooks, manuals, etc.
- science—experiments
- science projects—handbooks, manuals, etc.
- science projects—juvenile literature

Small Business Start-Up: A Research Guide

If you are interested in starting a small business, chances are you will need to begin by writing a business plan. Information needs will include how to raise capital and finance your business, industry analysis, market research, advertising plans, and identifying trade shows and conventions within your industry. This guide introduces researchers to some of the basic informational sources on the topic. The terms and phrases listed in the subject headings below can be used to search for more materials in the library's catalog and research databases. If you need further assistance, please ask a librarian.

Planning and Research

Almanac of Business and Industrial Financial Ratios by Leo Troy. Englewood Cliffs, NJ: Prentice Hall, 1969– . Annual. ISSN: 07479107.
Provides financial ratios for companies in various asset ranges by SIC code. Data reported are based on aggregate numbers from IRS records.

Annual Statement Studies. Philadelphia: Robert Morris Associates, 1923– . Annual. ISSN: 15457699.
Contains composite financial data on various industries. Used to compare one company's performance relative to other companies in the same line of business.

Business Plans Handbook. Detroit: Gale Research, 1995– . Annual. ISSN: 10844473.
Compilation of actual business plans developed by small businesses throughout North America.

Industry Norms and Key Business Ratios. Murray Hill, NJ: Dun & Bradstreet, 1983– . Annual.
Includes financial norm and business ratio data developed from actual company income statements and balance sheets.

Small Business Sourcebook: The Entrepreneur's Resource. Detroit: Gale Research, 1983– . Annual. ISBN: 0787670154.
Lists associations, consultants, educational programs, franchisers, government agencies, reference works, statisticians, suppliers, trade shows, and venture capital firms for various types of small businesses.

Small Business State Profiles. Office of Advocacy, U.S. Small Business Administration. Available at *www.sba.gov/ADVO/stats/profiles/*.
Includes basic information on the small business economy such as small business income, industry composition, number of firms, women and minority-owned firms, job creation, and banking information.

Financing

"The Myth of Free Government Money: A Perennial and Pernicious Scam. Why do so many entrepreneurs believe in the Tooth Fairy?" *BusinessWeek,* January 11, 2000. Available at *www.businessweek.com/smallbiz/0001/sa000111.htm.*

SBA Financing Center
www.sba.gov/financing/
Financing section of U.S. Small Business Administration site includes information on its various loan programs, forms, studies, workshops, shareware, and other resources.

Top 100 Venture Capital Firms/Entrepreneur Magazine
www.entrepreneur.com/listings/vc100/0,5946,,00.html
Offers information about venture capital in addition to directory.

Venture Capital Resource Library
www.vfinance.com
Provides research, investment banking, brokerage, and trading services to more than 10,000 corporate, institutional, and private clients worldwide.

Organizations

American Small Businesses Association (ASBA)
www.asbaonline.org
Advocacy group representing small business owners. Provides paid members with group benefits such as insurance, human resource benefits, and discounted services.

National Association for the Self Employed (NASE)
www.nase.org
Provides a broad range of benefits and support for self-employed and micro-businesses (up to ten employees).

Service Corps of Retired Executives (SCORE)
www.score.org
Nonprofit association funded by the SBA and dedicated to providing entrepreneurs with free, confidential face-to-face and e-mail business counseling.

Small Business Council of America (SBCA)
www.sbca.net
National organization representing the federal income and estate tax, pension, and other benefit interests of privately held and family owned businesses.

U.S. Small Business Administration
www.sbaonline.sba.gov
Its mission is to maintain and strengthen the nation's economy by aiding, counseling, assisting, and protecting the interests of small businesses and by helping families and businesses recover from national disasters.

Magazines

Entrepreneur. Santa Monica, CA: Chase Revel, 1978– . Monthly.
www.entrepreneur.com
Monthly magazine and informative Web site that offers suggestions on everything from finding your niche to marketing, advertising, and growing your business.

Inc.: The Magazine for Growing Companies. Boston: United Marine, 1979– . Monthly.
www.inc.com
Delivers advice, tools, and services, to help business owners and CEOs start, run, and grow their businesses more successfully.

Subject Headings

- business enterprises—United States—finance—statistics
- business planning—United States—handbooks, manuals, etc.
- financial statements—United States—statistics—periodicals
- self-employed—periodicals
- small business—information services—United States
- small business—United States—bibliography
- small business—United States—handbooks, manuals, etc.
- small business investment companies—directories
- venture capital—directories

Social Welfare: A Research Guide

Researchers looking for information about social welfare may be interested in research materials that discuss how such social welfare assists people with drug addiction, mental illness, poor housing, and provide counseling to neglected or abused children or the elderly—or the impact of such ills on society. Researchers may also be seeking contact information for agencies that provide the services that counsel, support, or provide guidance to people in need. This research guide will provide some of the basic resources to help you get started. The terms and phrases listed in the subject headings below can be used to search for more materials in the library's catalog and research databases. If you need further assistance, please ask a librarian.

Reference Books

Encyclopedia of Social Work, 19th edition. 3 vols. Washington, DC: National Association of Social Workers, 1997. ISSN: 00710237.
Provides an overview of more than 240 social welfare topics and definitions for social welfare and related terminology. Useful bibliographies at the end of each article. Includes biographies of prominent persons in the social work field, the NASW code of ethics, a chronology of the field, and acronyms.

The Social Work Dictionary, 4th ed. by Robert L. Barker. Washington, DC: National Association of Social Workers, 1999. ISBN: 0871012987.
Definitions for more than 5,000 terms, concepts, organizations, laws and historic events in the areas of social work administration, research, policy development, planning, community organization, human growth and development, health and mental health, clinical theory and practice. Also contains NASW Code of Ethics, state boards regulating social work.

Social Workers' Desk Reference edited by Albert R. Roberts and Gilbert J. Greene. New York: Oxford University Press, 2002. ISBN: 019514211X.
Covers the full spectrum of social work practice, focusing on the needs of frontline practitioners in private, nonprofit, and public settings (case managers, clinical social workers, supervisors, and administrators). It provides key tools and resources, such as best practices, program evaluations, step-by-step treatment plans, and validated assessment scales.

Periodical Indexes

Social Service Abstracts. Bethesda, MD: Cambridge Scientific Abstracts. Monthly. ISSN: 03094693.
www.csa2.com
Subscription service provides selective coverage of more than 1500 journals in the human services and related topics. Monthly publication.

Social Work Abstracts. Washington, DC: National Association of Social Workers, 1965– . Quarterly. ISSN: 10705317.
Index with abstracts to approximately 100 journals in social work and related subjects as well as dissertations in social work. Also available in electronic version.

Sociological Abstracts. San Diego: Sociological Abstracts, 1953– . Bimonthly. ISSN: 00380202.
Covers the field of sociology including such topics as the family, sociology of health care, social problems, human services delivery, social policy.

Directories

NASW Register of Clinical Social Workers. Silver Spring, MD: National Association of Social Workers, 1976– . Biennial. ISSN: 02770695.
Lists clinical social workers by name, address, agency, education, license, practice code, or area of specialization.

National Directory of Children, Youth & Families Services. Longmont, CO: Marion L. Peterson, 1979– . Annual. ISSN: 1072902X.

Addresses, phone numbers, and selected staff of public and private agencies throughout the United States.

Public Human Services Directory. Washington, DC: American Public Welfare Association, 1940– . Annual. ISSN: 15211320.

State and county services throughout the United States and Canada are listed with selected staff, addresses and phone numbers. Also includes some information on the organization of human services within each state or province.

Internet Resources

Social Welfare, Directory of Web Sites Created by the University at Albany (SUNY)
http://library.albany.edu/subject/socwork.htm

Topics include advocacy; children, youth and families; gerontology; government resources; mental health and therapy; social policy; Substance abuse; support services; statistical sources; professional associations.

World Wide Web Resources for Social Workers
www.nyu.edu/socialwork/wwwrsw/

Database of government, higher education, journal/newsletter, association, reference, and topical sites for social workers.

Subject Headings

- human services
- public welfare
- social service—dictionaries
- social service—guides, manuals, etc.
- social service—handbooks, manuals, etc.
- social service—periodicals
- social service—United States—dictionaries
- social welfare
- social welfare—directory
- social welfare—encyclopedias
- social work
- social work—directory
- social work—encyclopedias

Space Exploration: A Research Guide

Space exploration began in the 1950s and it was during this time that the National Aeronautics and Space Administration (NASA) was established on October 1, 1958. Its successor was the National Advisory Committee for Aeronautics, which was created in 1915. In earlier years, the space race between the United States and the Soviet Union fueled much of the advances we see today—its successes and failures. The resources below will assist researchers and students find out more about this fascinating topic. The terms and phrases listed in the subject headings below can be used to search for more materials in the library's catalog and research databases. If you need further assistance, please ask a librarian.

Books

The Cambridge Encyclopedia of Space by Fernand Verger, et al. New York: Cambridge University Press, 2002. ISBN: 0521773008.
Provides information on the occupation and use of space for scientific, industrial, commercial, technical, or military purposes. Includes many illustrations, maps, and graphs.

The Complete Book of Spaceflight: From Apollo 1 to Zero Gravity by David J. Darling. New York: Wiley, 2003. ISBN: 0471056499.
Includes 3,000 entries covering cultural development of spaceflight, including its history, science, and technology, the people involved, spaceflight's role in popular culture, and much more.

Journey into Space: The First Three Decades of Space Exploration by Bruce C. Murray. New York: W. W. Norton, 1989. ISBN: 0393026752.
Murray provides a personal account of the nation's planetary exploration program from the "golden age" of the 1960s and early 1970s.

Space Exploration by Christopher Mari. New York: H. W. Wilson, 1999. ISBN: 082420963X.
Covers John Glenn's return to space, the exploration of Mars, the International Space Station, private enterprise and space exploration, and new technologies and discoveries.

Magazine Articles

"Kennedy's Space Policy Reconsidered: A Post-Cold War Perspective," by Roger D. Launius. *Air Power History*, Winter 2003, v50, i4, p16.

"Return to Mars: How NASA plans to use its two rovers to roam the Red Planet looking for signs of Martian life," by Jeffrey Kluger. *Time*, January 19, 2004, v163, i3, Special Issue p52.

"So, You Want to Go to Mars: Sorry, you can't pack your toothbrush yet. But in January 2004, two NASA rovers will bring you closer than ever to our planet next door," by Mona Chiang. *Science World*, December 8, 2003, v60, i6, p8.

Web Sites

Johnson Space Center
www.jsc.nasa.gov
The Johnson Space Center was established in 1961 in Houston, Texas. This Web site includes everything from the early Gemini, Apollo, and Skylab projects to the Space Shuttle and International Space Station Programs.

Kennedy Space Center
www.ksc.nasa.gov
NASA's launch headquarters 45 minutes from Orlando.

Key Documents in the History of Space Policy
www.hq.nasa.gov/office/pao/History/spdocs.html
Full-text documents from the National Aeronautics and Space Act of 1958 to President George W. Bush's Space Exploration Program (2004).

NASA Multimedia Gallery
www.nasa.gov/multimedia/highlights/index.html
Find images and audio files about space exploration.

NASA Space Shuttle Launches
http://science.ksc.nasa.gov/shuttle/missions/missions.html
Provides details on all shuttle launches from 1981 to 2003.

Stennis Space Center
www.ssc.nasa.gov
Responsible for NASA's rocket propulsion testing and for partnering with industry to develop and implement remote sensing technology.

Organizations

The Mars Society
www.marssociety.org
Promotes the vision of pioneering Mars and assists in support of mars exploration programs.

President's Commission on Moon, Mars and Beyond
www.moontomars.org
President's commission on implementation of United States space exploration policy.

Subject Headings

- astronautics—encyclopedias
- outer space—exploration
- outer space—exploration—encyclopedias
- rocketry—encyclopedias
- space stations

Space Shuttle: A Research Guide

A space shuttle is a reusable space aircraft that takes off like a rocket and lands like an airplane. On April 12, 1981, the United States first space shuttle called Columbia blasted off. The shuttle was the first reusable spaceship and the first spacecraft able to land at an ordinary airfield. Since then there has been two space shuttle tragedies including the space shuttle Challenger that was destroyed shortly after its launch on January 28, 1986 and Columbia that was destroyed in an accident in midair on February 1, 2003. This guide introduces researchers to some of the basic informational sources on the topic. The terms and phrases listed in the subject headings below can be used to search for more materials in the library's catalog and research databases. If you need further assistance, please ask a librarian.

Books

Astronaut: Living in Space by Kate Hayden. New York: Dorling Kindersley, 2000. ISBN: 078945422X.
Describes the various steps necessary for an astronaut to prepare for a flight into space and follows astronaut Linda Gardner as she trains for her space shuttle mission.

The Challenger Disaster: Tragic Space Flight by Carmen Bredeson. Berkeley Heights, NJ: Enslow, 1999. ISBN: 0766012220.
Describes the events surrounding the explosion of the Challenger shuttle in 1986, the investigation of this disaster, and the seven astronauts who died.

Floating in Space by Franklyn Mansfield Branley and True Kelley. New York: HarperCollins, 1998. ISBN: 0060254327.
Examines life aboard a space shuttle, describing how astronauts deal with weightlessness, how they eat and exercise, and some of the work that they do.

On the Shuttle: Eight Days in Space by Barbara Bondar and Roberta Lynn Bondar. Toronto: Greey de Pencier Books, 1999. ISBN: 1895688124.
Day-by-day record of the Discovery mission.

The Space Shuttle by Allison Lassieur. New York: Children's Press, 2000. ISBN: 0516220039.
Describes the beginnings of the space shuttle program, the layout of a shuttle, a day's activities in space, the Challenger disaster, and the role of a shuttle in building a space station.

Biographies

Christa McAuliffe: A Space Biography by Laura S. Jeffrey. Springfield, NJ: Enslow, 1998. ISBN: 0894909762.
A biography of the school teacher turned astronaut whose life was tragically ended when the space shuttle Challenger exploded just after liftoff.

Shannon Lucid: Space Ambassador by. Carmen Bredeson. Brookfield, CT: Millbrook Press, 1998. ISBN: 0761304061.
Chronicles the life of the astronaut from her childhood in Oklahoma through her various space shuttle missions to her six months aboard the Mir space station.

Space Challenger: The Story of Guion Bluford by James Haskins and Kathleen Benson. Minneapolis: Carolrhoda Books, 1984. ISBN: 0876142595.
A biography of Guy Bluford, the first black American in space, part of the crew of the space shuttle "Challenger" on its August 1983 flight.

Magazine Articles

"Building a better shuttle: NASA turns to new designs and materials," by Ron Cowen. *Science News,* April 5, 2003. p215. Available at *www.findarticles.com/cf_0/m1200/14_163/100110924/p1/article.jhtml.*

"Why did the space shuttle burn up? (Columbia Disaster)," by Ron Cowen. *Science News*, February 8, 2003. p83. Available at *www.findarticles.com/cf_0/m1200/6_163/97997808/p1/article.jhtml.*

Web Sites

How Space Shuttles Work
http://science.howstuffworks.com/space-shuttle.htm
Describes how space shuttles work.

Kennedy Space Center: Space Shuttle
www.ksc.nasa.gov/shuttle/index.htm
This site includes Shuttle launch schedule, status reports, shuttle archives, shuttle factoids, shuttle landings, a reference manual.

Lost: Space Shuttle Columbia
www.cnn.com/SPECIALS/2003/shuttle/
CNN reports on the devastating loss of the space shuttle Columbia.

NASA Human Space Flight
http://spaceflight.nasa.gov
Behind-the-scenes, past and future mission information.

NASA Kids: The Space Shuttle
http://kids.msfc.nasa.gov/Rockets/shuttle.asp
Learn how did the shuttles get their names, what the different parts of the shuttle are, and look at some great shuttle pictures.

Space Shuttle Components
www.unitedspacealliance.com/shuttle/
Overview of the Space Shuttle and a clickable map of the shuttle but with more detail than the site above.

The World Almanac for Kids Online: Space Shuttle
www.worldalmanacforkids.com/explore/space/spaceshuttle.html
Brief overview of the development and history of the space shuttle program.

Subject Headings

- astronauts
- Challenger (spacecraft)
- Columbia (spacecraft)
- extraterrestrial environment—popular works
- life support systems (space environment)
- space exploration
- space flight—physiological effect
- space flight—popular works
- space shuttles

Standards and Specifications: A Research Guide

Standards and specifications mandated by organizations within certain industries undergo revision and change periodically. Standards regulate many industries and lines of work such as construction, agriculture, metals, and fire and building codes. Researchers who require access to this information will find a variety of resources listed below to help them with their search. The terms and phrases listed in the subject headings below can be used to search for more materials in the library's catalog and research databases. If you need further assistance, please ask a librarian.

Directories

Federal Information Processing Standards Index. Springfield, VA: U.S. Department of Commerce, National Institute of Standards and Technology, Computer Systems Laboratory, 1968– . Annual. ISSN: 00831816.
NIST develops FIPS for Federal computer systems when there are compelling Federal government requirements, such as for security and interoperability, and there are no acceptable industry standards or solutions.

Global Engineering Documents
http://global.ihs.com
Commercial organization with database searchable by document number, title words and date range, and results indicating price and availability.

Index and Directory of Industry Standards. Englewood, CO: Information Handling Services, 1983– . Annual. ISBN: 0898470080.
Indexes publications of the United States and international standards organizations and provides contact information for issuing associations.

Index of Federal Specifications, Standards and Commercial Item Descriptions. Washington, DC: General Services Administration, Federal Supply Services, 1952– . Annual. ISSN: 01989138.
Also available at *http://apps.fss.gsa.gov/pub/fedspecs/index.cfm.*
Provides access to federal standards and specifications.

Index of Specifications and Standards. Department of Defense (DODISS). Washington, DC: Standardization Division, Armed Forces Supply Support Center, 1979– . Annual. ISSN: 03638464. Available at *http://dodssp.daps.mil.*
Listing of United States Department of Defense standardization documents.

KWIC Index of International Standards. Geneva: International Organization for Standardization, 1983– . Biennial. ISBN: 9267101021.
Lists key word access to standards of international organizations.

NSSN: National Standards Systems Network
www.nssn.org
Database of standards searchable by title words or document number to identify standards and issuing organizations.

Organizations

ANSI (American National Standards Institute)
http://web.ansi.org
United States source for ISO documents. Includes reference library links to national standards bodies, international and regional standards organizations.

ASAE (American Society of Agricultural Engineers)
www.asae.org
Provides engineering standards applicable to agricultural, food, and biological systems.

ASHRAE (American Society of Heating, Refrigerating and Air Conditioning Engineers)
www.ashrae.org
Develops standards concerned with refrigeration processes and the design and maintenance of indoor environments.

ASTM (American Society for Testing and Materials)
www.astm.org/cgi-bin/SoftCart.exe/index.shtml?E+mystore
Standards covering metals, petroleum, construction, and the environment.

FASAB (Federal Accounting Standards Advisory Board)
www.fasab.gov
Develops federal accounting standards.

IES (Illuminating Engineering Society)
www.iesna.org
Lighting industry resource provides ordering information.

ISO Online
www.ico.ch
Source of ISO 9000, ISO 14000, and more than 14,000 International Standards for business, government, and society.

NFPA (National Fire Protection Association)
www.nfpa.org/Codes/index.asp
Develops, publishes, and disseminates codes and standards intended to minimize the possibility and effects of fire and other risks.

NISO (National Information Standards Organization)
www.niso.org/index.html
Develops standards, methods, materials, or practices for libraries, bibliographic and information services, and publishers.

NIST (National Institute of Standards and Technology)
www.nist.gov
Government agency working with industry on the development and applications of technology, measurements, and standards.

NSSN: National Standards System Network
www.nssn.org
Bibliographic information for more than 225,000 standards, searchable by title words or document numbers.

SAE (Society of Automotive Engineers)
www.sae.org
Resource for standards development, events, and technical information and expertise used in designing, building, maintaining, and operating self-propelled vehicles for use on land or sea, in air, or space.

UL (Underwriters Laboratories)
www.ul.com
Develops standards for safety.

Subject Headings

- industrial safety—United States—standards
- ISO 9000 series standards
- quality assurance—standards
- quality control—standards
- specifications
- standardization

Summer Olympics: A Research Guide

The materials listed below will entertain and inform you about the Summer Olympics. With a history that can be traced back to ancient Greece, the topic is both interesting and exciting. Olympian athletes dream of winning medals while representing their country and viewers cheer them on in excitement. Summer sports include track and field events, swimming, diving, fencing, and more. Additional information can be found on Olympians by searching their name in the library catalog or your favorite search engine. You can also search for magazine and newspaper articles. The terms and phrases listed in the subject headings below can be used to search for more materials in the library's catalog and research databases. If you need further assistance, please ask a librarian.

Books

Great Summer Olympic Moments by Nathan Aaseng. Minneapolis: Lerner, 1990. ISBN: 0822515369.
Discusses memorable athletic performances at the summer Olympics.

Hand Games by Mario Mariotti. Brooklyn, NY: Kane/Miller, 1992. ISBN: 091629143X.
Photographs of hands painted, decorated, and formed to resemble the players in various sports of the Summer Olympics.

The Olympic Summer Games by Caroline Arnold. New York: F. Watts, 1991. ISBN: 0531200523.
Discusses the history and organization of the Olympics, describing the individual sporting events of the summer Olympics.

The Summer Olympics by Bob Knotts. New York: Children's Press, 2000. ISBN: 0516210645.
Describes the history, ideals, events, and heroes of the Olympic Games, with an emphasis on the Summer Olympics.

Web Sites

Athens, Greece 2004
www.athens.olympic.org/Page/default.asp?la=2
Official site of the Athens 2004 Summer Olympics.

Beijing, China 2008
www.beijing-2008.org/new_olympic/eolympic/eindex.shtm
Official site of the Summer Olympic Games in Beijing.

International Olympic Committee
www.olympic.org/
Official site of the International Olympic Committee (IOC).

United States Olympic Committee
www.olympic-usa.org
United States Committee that organizes and promotes the games for team USA.

United States Olympic Team
www.usolympicteam.com/
Biographies of current and past athletes of the Olympics.

Olympic History

Coubertin Museum
www.museum.olympic.org/e/gallery/permanent/cou_bio_e.html
Museum that highlights the life and work of Pierre Frédy, Baron de Coubertin, the father of the modern Olympics.

Coubertin's Olympics: How the Games Began by Davida Kristy. Minneapolis: Lerner, 1995. ISBN: 0822533278.
Traces the history of modern day Olympics back to Pierre de Coubertin (1863–1937).

History of Olympic Mascots
www.collectors.olympic.org/e/fimo/fimo_mascots_e.html
Everything you every wanted to know about Olympic mascots, from "Schuss," the first unofficial Olympic Mascot in Grenoble in 1968 and "Waldi the Dachshund," the first official mascot in Munich in 1972, to the present day.

The Nazi Olympics: Berlin 1936 by Susan D. Bachrach. Boston: Little, Brown, 2000. ISBN: 0316070866.
Recounts the story of the Olympics held in Berlin in 1936.

The Olympic Festival in Antiquity
http://sunsite.icm.edu.pl/olympics/classical/intro.html
Learn about the people, places, rules, and games of the Ancient Olympics.

Olympic Museum
www.museum.olympic.org
Learn about Coubertin and Olympic History, explore a library of Olympic images and sounds, and museum exhibits.

Perseus Project: The Olympic Games
www.perseus.tufts.edu/Olympics/
Compare ancient and modern Olympic sports, tour the site of Olympia as it looks today, and learn about the ancient Olympics.

The Story of the Olympics by Dave Anderson. New York: W. Morrow, 1996. ISBN: 0688129544.
History of the Olympics back to 776 B.C. Includes stories of particular events such as track and field, gymnastics, and speed skating.

Subject Headings

- athletes—biography—juvenile literature
- Olympics—encyclopedias
- Olympics—history
- Olympics—history—juvenile literature
- Olympics—juvenile literature
- Summer Olympics—encyclopedias
- Summer Olympics—encyclopedias—juvenile literature
- Summer Olympics—history
- Summer Olympics—history—juvenile literature

Terrorism: A Research Guide

For students writing papers about terrorism or for researchers that want to know more, there is a great number of informational sources available to them about terrorism. Coverage includes methods, impact, foreign relations, personal loss, and the various response to terrorism. The form of this information comes in pictures, charts, maps, Web sites, printed materials, and government documents. The terms and phrases listed in the subject headings below can be used to search for more materials in the library's catalog and research databases. If you need further assistance, please ask a librarian.

Reference Books

Encyclopedia of Terrorism by Harvey W. Kushner. Thousand Oaks, CA: Sage, 2003. ISBN: 0761924086.
Over 300 articles that cover such topics as Al-Qaeda, biological terrorism, extremism, Saddam Hussein, jihad, Zacarias Moussaoui, and suicide bombers. Includes chronology, photographs, maps, and charts.

U.S. Foreign Policy Since the Cold War by Richard Joseph Stein. Bronx, NY: H. W. Wilson, 2001. ISBN: 0824209931
Includes a time line of U.S. foreign policy since the Cold War, America's approach to foreign policy, European relations, military intervention, China and "rogue states," and terrorism.

Internet Sites

Presidential News and Speeches
www.whitehouse.gov/news/releases/2001/09/
Presidential news releases and speeches from September 2001.

Response to Terrorism
http://usinfo.state.gov/topical/pol/terror/
Includes in-depth daily chronologies from September to December 2001, the other chronicling "Significant Terrorist Incidents" through the years 1961–2001. From the Office of International Information Programs, U.S. Department of State.

September 11th Web Archive
http://web.archive.org/collections/sep11.html
View archived Web sites in the aftermath of the terrorist attacks on the World Trade Center.

Shattered: A Photo Essay
www.time.com/time/photoessays/shattered/index.html
A collection of photographs by photojournalist James Nachtwey, sponsored by *Time* magazine.

News Coverage

Newspaper Headlines from the Poynter Institute
www.poynter.org/Terrorism/gallery/Extra1.htm
Front pages from special editions published on Tuesday, September 11, 2001.

PBS Frontline: Hunting Bin Laden
www.pbs.org/wgbh/pages/frontline/shows/binladen/
A coproduction of *PBS Frontline* and *The New York Times*, this site offers video footage of an interview with Bin Laden, and other transcripts from Bin Laden interviews throughout the years.

Government Information

Countering the changing threat of international terrorism: Report of the National Commission on Terrorism: Hearing before the Committee on Foreign Relations, United States Senate, June 15, 2000.
http://purl.access.gpo.gov/GPO/LPS4710

National Security Archive presents summaries and background government information as the *September 11th Sourcebooks*
www.gwu.edu/~nsarchiv/NSAEBB/sept11/

The Taliban: Engagement or Confrontation? Hearing before the Committee on Foreign Relations, United States Senate, July 20, 2000.
http://purl.access.gpo.gov/GPO/LPS10607

Organizations

Center for Defense Information: Terrorism Project
www.cdi.org/terrorism/
Committed to independent research on the social, economic, environmental, political, and military components of global security.

The Department of Homeland Security
www.whitehouse.gov/homeland/
A department created to make America a safer place after the terrorist attacks of September 11, 2001. Site monitors the "Nationwide Threat Level" according to the Homeland Security Advisory system devised by the Bush Administration.

The Terrorism Research Center
www.terrorism.com
Founded in 1996, The Terrorism Research Center offers information about cyberterrorism, terrorism links to organized crime, and bioterrorism. In addition, terrorist and counterterrorist profiles can be found here.

Subject Headings

- bioterrorism
- cyberterrorism
- international offenses
- political crimes and offenses
- state-sponsored terrorism
- terrorism
- terrorism—encyclopedias
- terrorism—government policy—United States
- terrorism—Middle East
- terrorism—prevention
- terrorism—United States
- terrorists—biography—dictionaries

Test Preparation: A Research Guide

We all take tests at some time in our lives. Perhaps we need to take an exam to enter school or a higher level of education like college or graduate school, or even law school. Some of us must past exams to enter into a new occupation or continue on in our current careers for certification. Below you will find some books and Web sites that will help prepare you for your next exam. This guide introduces researchers to some of the basic informational sources on the topic. The terms and phrases listed in the subject headings below can be used to search for more materials in the library's catalog and research databases. If you need further assistance, please ask a librarian.

Web Sites

ACT Assessment
www.act.org/aap/index.html
This site details the ACT, which is designed to assess high school students' general educational development and their ability to complete college-level work.

College Board
www.collegeboard.com
Provides information on test sponsored by the College Board including the SAT (Scholastic Aptitude Test), PSAT/NMSQT (Preliminary SAT/National Merit Scholarship Qualifying Test), AP (Advanced Placement Program), and CLEP (College Level Examination Program).

Graduate Record Examinations (GRE)
www.gre.org
Helps to prepare you for taking the GRE.

Law School Admission Council
www.lsat.org
A nonprofit corporation whose members are 202 law schools in the United States and Canada. Administers the Law School Admission Test (LSAT).

LibrarySpot: Where to Prepare for the Big Test
www.libraryspot.com/features/testprepfeature.htm
Offers an overview of test preparation and links to further resources.

Number2.com
www.number2.com
Provides free practice tests for the ACT, SAT, GRE, and vocabulary builder.

PBS's Secrets of the SAT
www.pbs.org/wgbh/pages/frontline/shows/sats/
This report draws on the work of Nicholas Lemann and his five-year study of the SAT—*The Big Test: The Secret History of the American Meritocracy.*

Test Preparation Companies

Barron's Test Preparation
www.barronstestprep.com
Publishes, for a fee, both online and print preparatory examinations for PSAT, SAT, and ACT tests.

Kaplan Test Prep and Admissions
www.kaptest.com
Publishes, for a fee, both online and print preparatory examinations for college, graduate school, civil service and military careers, and occupations.

Peterson's
www.petersons.com
Publishes, for a fee, both online and print preparatory examinations for college, graduate school, civil service and military careers, and occupations. Also includes college and tuition research. Also publishes titles with the Arco imprint.

Princeton Review
www.princetonreview.com
Publishes, for a fee, both online and print preparatory examinations for college, graduate school, civil service and military careers, and occupations. Also includes college searches and test registration.

Test Categories

Below are just some of the test titles or categories that require testing.

- ACT (American College Test)
- AP (Advanced Placement Program)
- basic skills
- civil service
- CLEP (College Level Examination Program)
- correctional officer
- cosmetology
- EMS (Emergency Medical Services)
- ESL (English as a Second Language)
- firefighter
- GED (General Equivalency Degree)
- GMAT (Graduate Management Admission Test)
- GRE (Graduate Record Examinations)
- law enforcement
- LSAT (Law School Admission Test)
- MCAT (Medical College Admissions Test)
- military services
- nurse aide
- PSAT/NMSQT (Preliminary SAT/National Merit Scholarship Qualifying Test)
- police officer
- postal service
- real estate
- SAT (Scholastic Aptitude Test)
- SAT II Subject Tests
- skills improvement courses
- teacher certification
- technical and career college
- TOEFL (Test of English as a Foreign Language)
- U.S. citizenship
- USMLE (United States Medical Licensing Examination)

Subject Headings

- accounting—examinations, questions, etc.
- admission test
- civil service—United States—examinations
- computer networks—examinations—study guides
- entrance examination
- general educational development tests—study guides
- high school equivalency examinations—study guides
- postal service—United States—employees
- postal service—United States—examinations, questions, etc.
- study guides

Theater: A Research Guide

Whether you are researching historical information about playwrights, theories, traditions, or stage management and stage makeup this research guide lists many useful resource materials. Subject headings listed at the end of this guide can be searched in the library catalog locating even more materials on this topic. The materials below will help beginning researchers as well as those looking for more comprehensive coverage. The terms and phrases listed in the subject headings below can be used to search for more materials in the library's catalog and research databases. If you need further assistance, please ask a librarian.

General Works

The Cambridge Guide to Theatre edited by Martin Banham. New York: Cambridge University Press, 1988. ISBN: 0521434378.
Covers all major playwrights, works, traditions, theories, companies, venues and events; details the origins of popular theater tradition.

Contemporary Dramatists, 6th ed. Detroit: St. James Press, 1998. ISBN: 0912289627.
Entries for each dramatist in this work include biographical details, a critical essay, a works list, etc. Some also include a comment by the person listed. Entries are indexed by nationalities and the titles of plays.

History of the Theatre, 9th ed. edited by Oscar G. Brockett and Franklin J. Hildy. Boston: Allyn & Bacon, 2003. ISBN: 0205358780.
This general history of the theater provides overviews of aspects of European and American theater, as well as the theater of Asia, Africa, etc.

McGraw-Hill Encyclopedia of World Drama: An International Reference Work, 2nd ed. edited by Stanley Hochman. 5 vols. New York: McGraw-Hill, 1984. ISBN: 0070791694.
This illustrated reference includes biographies of dramatists, some summaries, and biographical and critical references. Volume 5 is a general index, and articles on theater companies provide additional information.

The Oxford Companion to American Theatre edited by Gerald Bordman. New York: Oxford University Press, 1992. ISBN: 0195072464.
The emphasis in this reference is on Broadway theater. Entries on plays include information, such as the producer, the length of the play's run, the original date of production, etc.

The Oxford Companion to the Theatre, 4th ed. edited by Phyllis Hartnoll. New York: Oxford University Press, 1983. ISBN: 0192115464.
Includes topics pertaining to both classical and contemporary theater. Included are entries on actresses, directors, techniques, and acting companies.

Play Sources and Production

Ottemiller's Index to Plays in Collections: An Author and Title Index to Plays Appearing in Collections Published Between 1900 and 1985, 7th ed., rev. and enl. edited by John H. Ottemiller, Billie M. Connor, and Helene G. Mochedlover. Metuchen, NJ: Scarecrow Press, 1988. ISBN: 0810820811.
This resource offers an author index to 6,548 titles of plays located in collections that were published mostly in the United States or England. Plays are indexed by title.

Play Index. New York: H. W. Wilson, 1949/1952– .
Items in this work are indexed by author, title, and subject. Entries are of English-language plays and translations and include brief plot summaries.

Stage Makeup, 9th ed. by Richard Corson and James Glavan. Boston: Allyn & Bacon, 2001. ISBN: 0136061532.

A variety of aspects concerning stage makeup are present in this work. Sections include: "Facial Anatomy," "Hair and Wigs," "Prosthetic Makeup," and "Relating the Makeup to the Character," etc.

Reviews and Criticism

American Drama Criticism: Interpretations, 1890–1977, 2nd ed. compiled by Floyd Eugene Eddleman. Hamden, CT: Shoe String Press, 1979. ISBN: 0208017135.

Contains references to interpretations of American plays found in books, monographs, and periodicals. It includes mostly works of United States dramatists with a few exceptions. Musical plays are also examined.

Critical Survey of Drama, 2nd rev. ed. edited by Frank N. Magill. 8 vols. Pasadena, CA: Salem Press, 2003. ISBN: 1587651025.

Contains biographical information about the authors and their works, including significant achievements. Critical essays on drama and its developments over the years are located in the last volume of this set.

Web Sites

Internet Broadway Database
www.ibdb.com

IBDB provides a comprehensive database of shows produced on Broadway, including all "title page" information about each production. IBDB also offers historical information about theaters and various statistics and fun facts related to Broadway.

Performing Arts in America 1875–1923
http://digital.nypl.org/lpa/nypl/lpa_home4.html

A database of 16,000 archival materials from the New York Public library. Includes newspaper clippings, composite photographs, music sheet samples featuring popular music, show tunes, jazz and dance music; photographs of theater, dance, and popular performance; and publicity posters and lobby cards.

Theatermania.com
www.theatermania.com

Lists productions of major cities including information on theaters, ticket price ranges, and show types.

Subject Headings

- actors—England—London—biography—dictionaries
- American drama
- Commonwealth drama
- drama—history and criticism—dictionaries
- English drama
- performing arts
- stage lighting
- stage machinery
- stage management
- theater—dictionaries
- theater—history—20th century
- theaters—stage setting and scenery
- theatrical makeup
- theatrical managers

Travel Information: A Research Guide

Create your custom dream vacation, search for travel bargains, or simply read a travel guide. Travel the world from the comfort of your home computer. The materials listed below allow you to interact with other travelers who have "been there, done that" and also allow you to purchase your trip online. This guide introduces researchers to some of the basic informational sources on the topic. The terms and phrases listed in the subject headings below can be used to search for more materials in the library's catalog and research databases. If you need further assistance, please ask a librarian.

Guidebooks

Many guidebooks are available titled by travel destination and offer listings, ratings, and evaluations of accommodations, restaurants, and attractions. Some of the titles include Fodor's, Frommer's, Lonely Planet, and AAA guides. Many have accompanying Web sites like those listed below.

Fodor's: *www.fodors.com*
Frommer's: *www.frommers.com*
Let's Go!: *www.letsgo.com*
Lonely Planet: *www.lonelyplanet.com*
Rough Guides: *http://travel.roughguides.com*

Magazines

Arthur Frommer's Budget Travel. New York: Arthur Frommer Magazines, 1998– . Bimonthly. ISSN: 15215210.
Provides indispensable information for the budget traveler.

Condé Nast's Traveler. New York: Condé Nast, 1966– . Monthly. ISSN: 08939683.
www.concierge.com/cntraveler/
Focus to provide the experienced traveler an array of diverse travel experiences encompassing art, architecture, fashion, culture, cuisine, and shopping.

National Geographic Traveler. Washington, DC: National Geographic Society, 1984– . Bimonthly. ISSN: 07470932.
Focuses on domestic and foreign destinations, personal travel reflections, food and restaurants, great places to stay, photography, trends, adventure, ecotourism, road trips, cultural events, and travelers.

Travel Holiday. New York: Hachette Filipacchi, 1911– . Monthly. ISSN: 0199025X.
www.travelholiday.com
Provides tips on travel, freebies, best deals, and destinations.

Travel & Leisure. New York, American Express, 1941– . Monthly. ISSN: 00412007.
www.travelandleisure.com
Provides useful travel tips, off-the-beaten-path itineraries, the best hotels and restaurants, and recommendations for the next great place to visit—and how to get there.

Online Booking

In addition to registering directly at an airline, hotel, or rental car Web site, the following Web sites provide a full array of online booking services:

Expedia.com
www.expedia.com
Flights, hotels, cars, package deals, and cruises. Use the Fare Tracker to receive e-mail when your cities of interest go on sale.

Lowestfare.com
www.lowestfare.com
Offers discount fares on hotels, airfare, cruises, and travel packages. Also offers last minute specials, online booking, and toll-free customer service telephone number.

Onetravel.com
www.onetravel.com
Search for low airfares, discount hotel deals, car rental specials, and vacation packages and specials.

Travelocity.com
www.travelocity.com
24-hour customer service via toll-free telephone line. Provides best deals on lodging, flights, car rentals, rail passes, vacation packages, and cruises. Offers Farewatcher service via e-mail. E-mail notification when your preferred destinations goes on sale.

Web Sites

Airport and City Code Converter
http://codes.managementreporting.com
Have a three-letter airport code and you're not sure of its location? Enter either the letter code to find out the location, current weather, and brief statistics on the airport.

BedandBreakfast.com
www.bedandbreakfast.com
Search and view thousands of bed and breakfasts from around the United States. Includes pictures, descriptions, rates, and online booking information.

CDC Travelers' Health
www.cdc.gov/travel/
Outlines potential hazards by destination. Lists known diseases and required vaccinations. Also offers tips on traveling with children outside of the United States, cruise and air travel, food and water precautions.

Flightarrivals.com
www.flightarrivals.com
Independent source of information for arrivals, departures, and delays for U.S. and Canadian flights.

Great Outdoor Recreation Pages
www.gorp.com
Provides information on outdoor and active travel including hiking, birding, climbing, fishing, and much more. Find dude ranches and book custom vacations.

Travel Warnings and Consular Information Sheets
http://travel.state.gov/travel_warnings.html
View travel warnings, public announcements relating to terrorist threats, and consular information sheets. You will find the following information: location of the U.S. Embassy or Consulate in the subject country, unusual immigration practices, health conditions, minor political disturbances, unusual currency and entry regulations, crime and security information, and drug penalties.

Virtual Tourist
www.vtourist.com
Provides tourist information from first-hand visitors who have traveled and documented their trips on these pages. Currency, time zone, and metric converters included.

Subject Headings

- camp sites, facilities, etc.—Canada—directories
- camp sites, facilities, etc.—United States—directories
- hotels—guidebooks
- hotels—pictorial works
- travel
- United States—description and travel
- United States—guidebooks

U.S. Civil Rights: A Research Guide

Civil rights are the freedoms that a person may have as a member of a community, state, or nation. They include freedom of speech and of the press, freedom of religion, the right to own property, and the right to receive fair and equal treatment from government, other persons, and private groups. The United States Supreme Court's decisions often defines civil rights by weighing individual rights against the rights of society in general. This guide introduces the researcher to some of the basic informational sources on the topic. The terms and phrases listed in the subject headings below can be used to search for more materials in the library's catalog and research databases. If you need further assistance, please ask a librarian.

Reference Books

The ABC-CLIO Companion to the Civil Rights Movement by Mark Grossman. Santa Barbara, CA: ABC-CLIO, 1993. ISBN: 0874366968.
Concise alphabetical entries cover the movement's major issues, landmark court decisions, organizations, and key concepts, terms, and events.

Civil Rights in the United States edited by Waldo E. Martin, Jr. and Patricia Sullivan. 2 vols. New York: Macmillan Reference USA, 2000.
Covers civil rights issues such as its sources and movements, including the Bill of Rights, as well as critical developments such as the African-American civil rights movement of post-World War II in America.

The Encyclopedia of Civil Rights in America edited by David Bradley and Shelley Fisher Fishkin. 3 vols. Armonk, NY: Sharpe Reference, 1998. ISBN: 0765680009.
Covers important issues and people related to civil rights in America including: Abernathy, Ralph to Equal protection clause (vol. 1); equal rights amendment to Presidency, U.S. (vol. 2), and Price, U.S. vs. to Zoot suit riots (vol.3).

My Soul is a Witness: A Chronology of the Civil Rights Era in the United States, 1954–1965 by Bettye Collier-Thomas and V. P. Franklin. New York: Henry Holt, 2000. ISBN: 0805047697.
Includes more than 2,500 entries that illustrate the civil rights era through the thousands of people, places, and events that the Civil Rights Movement encompassed in the areas of employment, public accommodations, housing, voting rights, religion, entertainment, sports, and the military. Based on articles and reports found in *The New York Times*, *Jet* magazine, and the *Southern School News*.

Biographies

American Women Civil Rights Activists: Biobibliographies of 68 Leaders, 1825–1992 by Gayle J. Hardy. Jefferson, NC: McFarland, 1993. ISBN: 0899507735.
Includes women who have worked passionately for civil rights in the United States such as Elizabeth Blackwell, Mary L. Bonney, Kate Barnard, Mariana Bracetti, and Amelia Stone Quinton, to contemporary figures such as Iola M. Pohocsucut Hayden, Rosa Parks, Angela Davis, and Shirley Chisholm.

Freedom's Daughters: The Unsung Heroines of the Civil Rights Movement from 1830 to 1970 by Lynne Olson. New York: Charles Scribner's Sons, 2001. ISBN: 0684850125.
Details the lives of the women who led the fight against lynching (Ida Wells), who organized the first sit-in at a lunch counter (Pauli Murray), who launched the Montgomery bus boycott (Jo Ann Robinson), and who were instrumental in turning the sixties civil rights crusade into a mass movement.

Web Sites

"Civil Rights Movement" in *Gale Encyclopedia of Popular Culture* by Leonard N. Moore.
www.findarticles.com/cf_0/g1epc/tov/2419100257/p1/article.jhtml
Provides a topical overview of the civil rights movement.

Civilrights.org
www.civilrights.org
Provides relevant and up-to-the minute civil rights news and information. A collaboration of the Leadership Conference on Civil Rights and the Leadership Conference on Civil Rights Education Fund.

Historical Publications of the United States Commission on Civil Rights
www.law.umaryland.edu/edocs/usccr/html%20files/usccrhp.asp
Provides access to hundreds of documents produced by the United States Commission on Civil Rights. Hosted by the Thurgood Marshall Law Library.

We Shall Overcome: Historic Places of the Civil Rights Movement
www.cr.nps.gov/nr/travel/civilrights/index.htm
Produced by the National Park Service (NPS), U.S. Department of the Interior, in cooperation with the Federal Highway Administration, U.S. Department of Transportation, and the National Conference of State Historic Preservation Officers (NCSHPO).

Organizations

American Civil Liberties Union
www.aclu.org
Aims to defend and preserve the individual rights and liberties guaranteed to every person in this country by the Constitution and laws of the United States.

Human and Civil Rights Organizations of America
www.hcr.org
A nonprofit federation that screens and certifies high quality national charities working to protect human rights and promote the liberty of all people.

Subject Headings

- African Americans—civil rights—encyclopedias
- African Americans—civil rights—history
- African Americans—history—1877–1964—encyclopedias
- civil rights—United States
- civil rights movements
- minorities—civil rights—United States
- women civil rights workers

U.S. Civil War: A Research Guide

Researchers of the U.S. Civil War (1861–1865) will find a great deal of information to quickly begin their research from the materials listed below. The war between the Northern states (the Union) and the Southern States (the Confederates) is also called the War of the Rebellion, War of Secession, and the War for Southern Independence. Students will find that a great deal of information awaits them once they have referenced the materials listed within this guide. Much of the material, in addition to text, are pictorial works, atlases, statistics, diaries, and cover the causes of the war, diplomatic relations, involvement of African-Americans, medical issues, prison, art, and music. The terms and phrases listed in the subject headings below can be used to search for more materials in the library's catalog and research databases. If you need further assistance, please ask a librarian.

Books for Children

American Civil War Almanac by Kevin Hillstrom, Laurie Collier Hillstrom, and Lawrence W. Baker. Detroit: UXL, 2000. ISBN: 0787638234.
This book presents a comprehensive overview of the war in chronological order starting with issues leading up to the war and ending with Reconstruction. Other notable chapters cover the roles of women and blacks in the war. A time line, glossaries, and sources for further reading are also included.

American Civil War Biographies by Kevin Hillstrom, Laurie Collier Hillstrom, and Lawrence W. Baker. Detroit: UXL, 2000. ISBN: 078763820X.
This set covers the life stories of sixty individuals important in the Civil War. The entries are in alphabetical order and include pictures, additional sources, and a time line.

American Civil War Primary Sources by Kevin Hillstrom, Laurie Collier Hillstrom, and Lawrence W. Baker. Detroit: UXL, 2000. ISBN: 0787638242.
This volume includes speeches, letters, diaries, and photographs by people who lived during the war. Each entry has an explanatory introduction and sources for additional reading.

Civil War A to Z: A Young Readers' Guide to over 100 People, Places, and Points of Importance by Norm Bolotin. New York: Dutton Children's Books, 2002. ISBN: 0525462686.
This book contains alphabetically arranged short entries on 100 important people, places, issues, and events of the Civil War.

Life in the North during the Civil War by Timothy L. Biel. San Diego: Lucent Books, 1997. ISBN: 1560063343.
This book tells about life in the army, in the city, and in the country and about the racial and political tensions present in northern society.

Life in the South During the Civil War by James P. Reger. San Diego: Lucent Books, 1997. ISBN: 1560063335.
The lifestyle of plantation owners, slaves, mountain folk, and the nonslave holding middle class is covered in this book, as well as the effect of total warfare on the south.

Songs and Stories of the Civil War by Jerry Silverman. Brookfield, CT: Twenty-First Century Books, 2002. ISBN: 0761323058.
This book tells the story of the war through the music of the time and includes lyrics and sheet music for each song.

Reference Books

The Civil War Day by Day: An Almanac, 1861–1865 by E. B. Long and Barbara Long. New York: Da Capo Press, 1985, 1971. ISBN: 0306802554.
Important events on a daily basis from President Lincoln's election into early 1866.

Encyclopedia of the American Civil War: A Political, Social, and Military History by David Stephen Heidler, Jeanne T. Heidler, and David J. Coles. Santa Barbara, CA: ABC-CLIO, 2000. ISBN: 1576070662.

This set covers political, social, and military history in entries arranged in alphabetical order. References for further reading as well as pictures, maps, a chronology of events, and many primary sources are included.

Historical Times Illustrated Encyclopedia of the Civil War edited by Patricia L. Faust and Norman C. Delaney. New York: Harper & Row, 1986. ISBN: 0061812617.

This book includes more than 2,000 entries ranging from several lines to several pages. It covers military events, campaigns, battles, political, economic and social developments, and biographies.

The Photographic History of the Civil War: Thousands of Scenes Photographed 1861–65 by Francis Trevelyan Miller and Robert S. Lanier. 10 vols. New York: Review of Reviews, 1911.

If you need pictures, this old multivolume set is a gold mine. Vol 1: The opening battles; vol. 2: 2 years of grim war; vol 3: The decisive battles; vol. 4: The cavalry; vol. 5: Forts and artillery; vol. 6: The navies; vol. 7: Prisons and hospitals; vol. 8: Soldier life, secret service; vol. 9: Poetry and eloquence of Blue and Gray; vol. 10: Armies and leaders.

Web Sites

Civil War Maps
http://memory.loc.gov/ammem/gmdhtml/cwmhtml/cwmhome.html
Reconnaissance, sketch, coastal, and theater-of-war maps that depict troop activities and fortifications during the Civil War are included in this collection from the Library of Congress.

A Nation Divided: The U.S. Civil War 1861–1865
www.historyplace.com/civilwar/
This site has a time line with many printable pictures.

The U.S. Civil War Center
www.cwc.lsu.edu
This site offers a huge list of links to all kinds of Civil War topics.

Subject Headings

- United States—history—Civil War, 1861–1865—art and the war
- United States—history—Civil War, 1861–1865—chronology
- United States—history—Civil War, 1861–1865—dictionaries
- United States—history—Civil War, 1861–1865—encyclopedias
- United States—history—Civil War, 1861–1865—indexes
- United States—history—Civil War, 1861–1865—pictorial works
- United States—history—Civil War, 1861–1865—portraits

U.S. Congress: A Research Guide

As politics and legislation have an increasing affect upon our daily lives, the more voters are motivated to learn more about their elected officials, the congressional process by which legislation comes about, and how it impacts the future. Congressional information can include historical information on voting pattern and biographical information and it can encompass timely information on weekly schedules and committee activity. The Internet has made this information more accessible than ever. Many publications once available only in print are now distributed through the Internet for more accurate consumption and action. This guide introduces researchers to some of the basic informational sources on the topic. The terms and phrases listed in the subject headings below can be used to search for more materials in the library's catalog and research databases. If you need further assistance, please ask a librarian.

General Research and Reference

The Almanac of American Politics, 2004 by Michael Barone and Richard E. Cohen. Chicago: University of Chicago Press, 2003. ISBN: 089234105X.
Guide to the American political scene including important issues like how redistricting will alter American politics over the course of the next decade; photographs of all 535 members of Congress and the 50 governors. Also covers voting records on important legislation, including congressional vote ratings by *National Journal* and interest groups.

Congressional Roll Call 2003: A Chronology and Analysis of Votes in the House and Senate. Washington, DC: CQ Press, 2004.
Includes legislative summary of the session, voting studies that examine presidential support, party unity, coalitions, voting trends, and key votes throughout the year.

Congressional Quarterly Weekly Report. Washington, DC: Congressional Quarterly, 1948– . Weekly.
Covers congressional activities. Includes indexes, chronologies, bill titles, and numbers. Informative finding tool, unbiased, comprehensive roundup of Capitol Hill activity from the previous week.

CQ Almanac Plus. Washington, DC: Congressional Quarterly, 2001– . Annual. Continues CQ Almanac (1948–2000).
Detailed look at each major bill considered during the legislative year Also contains useful data-filled appendixes, including: key votes, vote studies, roll call votes, public laws, and texts of key speeches.

GPO Access: Congressional Materials
www.gpoaccess.gov/legislative.html
Includes a broad variety and in-depth collection of congressional publications including congressional calendars, committees, rules and procedures, Congressional Serial Set, and independent counsel reports.

Thomas: Legislative Information on the Internet
http://thomas.loc.gov
Created to provide Federal legislative information freely available to the Internet public. Includes a variety of information-rich databases some of which are listed below including legislation, roll call votes, and committee information.

Legislation:

Bill Summary and Status: 93rd through 108th Congresses (1973–present)
Bill Text: 101st–108th Congresses (1989–present)
Public Laws by Law Number: 93rd through 108th Congresses (1973–present)

Votes:

House Roll Call Votes: 101st Congress, 2nd through 108th (1990–present)
Senate Roll Call Votes: 101st Congress, 1st through 108th (1989–present)

United States House of Representatives
www.house.gov
Offers a complete collection of publications and information about the people who make up the U.S. House of Representatives, how to contact them, committees, memberships, assignments, legislation and voting records, and the history of the House.

United States Senate
www.senate.gov
Offers a complete collections of publications and information about the people who make up the U.S. Senate, how to contact them, committees, memberships, assignments, legislation and voting records, and the history of the Senate.

Biographical Information

Biographical Directory of the American Congress, 1774–1996. Mount Vernon, VA: Staff Directories, 1996. A more current version is available online at *http://bioguide.congress.gov/biosearch/biosearch.asp.*
Information about any senator, representative, vice president, or member of the Continental Congress.

Congressional Pictorial Directory. Washington, DC: Government Printing Office. For sale by the Supt. of Docs., Government Printing Office. Biennial. Also Available online at *www.access.gpo.gov/congress/108_pictorial/index.html.*
Includes black-and-white photographs of all congressional members.

Congressional Staff Directory by Anna L. Brownson and Charles Bruce Brownson. Indianapolis: Bobbs-Merrill, 1959– . Three times a year. ISSN: 05893178.
Lists members of Congress and their staff, committees and the legislation they work on, lists states and their districts. Includes multiple indexes.

Subject Headings

- government—United States—periodicals
- labor laws and legislation—United States—periodicals
- legislation—United States—periodicals
- United States Congress—biography
- United States Congress—directories

U.S. Constitution: A Research Guide

Now over 200 years old, the U.S. Constitution and its basic tenets remain the same and continue to be challenged through the legal system. During the drafting of the Constitution, the Constitutional Convention pointed at issues including how much power to allow the central government, how many representatives in Congress to allow each state, and how these representatives should be elected—directly by the people or by the state legislators. Learn more about how the delegates debated, and redrafted the articles of the new Constitution through the research materials listed below. The terms and phrases listed in the subject headings below can be used to search for more materials in the library's catalog and research databases. If you need further assistance, please ask a librarian.

Books

The Constitution and Its Amendments by Roger K. Newman. 4 vols. New York: Macmillan Reference USA, 1999. ISBN: 0028648587.
Provides a chronological history of the Constitution's seven articles and 27 amendments. Features include sidebar definitions and examples, photos, and political cartoons.

Constitutional Amendments, 1789 to the Present edited by Kris E. Palmer. Detroit: Gale Group, 2000. ISBN: 0787607827.
Lists amendments. Includes the full text of the amendment (including those proposed but unratified) and essays illustrating the social and political contexts.

Constitutional Amendments: From Freedom of Speech to Flag Burning by Tom Pendergast, Sara Pendergast, John Sousanis, and Elizabeth Shaw Grunow. 3 vols. Detroit: UXL, 2001. ISBN: 0787648655
Entries range in length from 10 to 15 pages and begin with the full text of all 27 amendments, followed by an essay on the social and political climate that gave rise to its proposal. Includes illustrations, sidebars, and a bibliography.

Constitutional Rights Sourcebook by Peter G. Renstrom. Santa Barbara, CA: ABC-CLIO, 1999. ISBN: 1576070611.
Provides overview, 200 comprehensive entries on concepts, court decisions, people, and organizations, bibliography, table of cases, and index.

Encyclopedia of the American Constitution edited by Leonard Williams Levy, Kenneth L. Karst, and Dennis J. Mahoney. 4 vols. New York: Macmillan, 1986. ISBN: 0029186102.
Articles on individual subjects, individuals, landmark law decisions, and brief bibliographies related to the U.S. Constitution. 1992 supplement.

Encyclopedia of Constitutional Amendments, Proposed Amendments, and Amending Issues, 1789–2002, 2nd ed. by John R. Vile. Santa Barbara, CA: ABC-CLIO, 2003. ISBN: 1851094288.
Covers 27 amendments, as well as notable proposed amendments, along with related issues, individuals, organizations, and Supreme Court decisions, among other topics.

Web Sites

Charters of Freedom, National Archives
www.archives.gov/national_archives_experience/constitution.html
View original constitution and learn more about the men who drafted it.

Colonial Hall
www.colonialhall.com
Provides biographies on the signers of the U.S. Constitution.

Explore the Constitution, National Constitution Center
www.constitutioncenter.org, select Explore the Constitution
Explores some of the guiding principles of the Constitution such as rule of law, separation of powers and checks and balances, federalism, judicial review, and individual rights.

To Form a More Perfect Union
http://lcweb2.loc.gov/ammem/bdsds/intro01.html
Narratives and searchable database of documents relating to the work of the Continental Congress and the drafting and ratification of the Constitution. From the Library of Congress.

Organizations

National Constitution Center
www.constitutioncenter.org
An independent, nonpartisan, and nonprofit organization dedicated to increasing public understanding of, and appreciation for, the Constitution, its history, and its contemporary relevance.

Subject Headings

- civil rights
- Constitutional amendments—United States
- Constitutional conventions
- Constitutional history—United States
- United States—Constitution
- United States—Constitution—1st–10th amendments
- United States—Constitution—amendments
- United States—Constitutional convention, 1787
- United States—Constitutional history
- United States—Constitutional law
- United States—Supreme Court

U.S. Income Tax Preparation: A Research Guide

Up against the tax deadline? This guide is designed to provide information and assistance in the income tax preparation process. There are many resources available with extensive information and advice about taxes, both in print and on the Internet. These resources provide a range of information, including tax basics, the latest tax law changes, IRS forms and publications, tips for reducing income tax, and tax planning advice. This guide introduces researchers to some of the basic informational sources on the topic. The terms and phrases listed in the subject headings below can be used to search for more materials in the library's catalog and research databases. If you need further assistance, please ask a librarian.

Books

The Complete Idiot's Guide to Doing Your Income Taxes by Gail Perry. New York: Alpha Books, 1996– . Annual. ISSN: 10987193.
Intended for the average taxpayer without a complex tax situation. It provides instructions for all aspects of the tax process from organizing tax materials to filing the completed return. Tax planning suggestions are also given.

Ernst and Young Tax Guide. New York: John Wiley & Sons, 1985– . Annual. ISSN: 1059809X.
Guide to the tax system produced by an accounting firm. Includes tax tips and complete information about how to handle investment income.

H&R Block Income Tax Guide. New York: Collier Books, 1986– . Annual. ISSN: 10549846.
Guide that includes a tax organizer, worksheets, and calendars to assist in the tax return preparation process.

Taxes for Dummies. Foster City, CA: IDG Books Worldwide, 1995– . Annual. ISSN: 15351130.
Intended for the average taxpayer without a complex tax situation. It starts with an overview of the U.S. tax system and provides detailed line-by-line instructions for preparing a tax return.

Web Sites

FedWorld Tax Form Search
www.fedworld.gov/taxsear.html
Locate archived IRS forms and publications.

Internal Revenue Service
www.irs.ustreas.gov
Contains statistics, taxpayer information, forms, publications, and more. Everything that a taxpayer needs to know about filing a tax return can be found here.

Microsoft: MoneyCentral—Taxes
http://moneycentral.msn.com/tax/home.asp
The Microsoft Network provides tax information, news, tips, a tax estimator, and tax planning guides.

PayrollTaxes.com
www.payroll-taxes.com
Source of information about payroll taxes. It provides the ability to calculate take-home pay.

Small Business/Self-Employed
www.irs.gov/businesses/small/index.html
Tax preparation for small businesses and the self-employed.

SmartMoney: Tax Guide
www.smartmoney.com/tax/
Articles written in plain language explain various tax concepts and how to reduce tax liability.

Tax and Accounting Sites Directory
www.taxsites.com
Directory containing links to numerous tax and accounting sites, including state-specific taxing authority sites.

Tax Problems Law
www.nolo.com/encyclopedia/tax_ency.html
The Nolo Press "Self-help law center" provides advice on audits, tax bills, late payments, and dealing with the IRS.

Tax Tips from Bankrate.com
www.bankrate.com/brm/itax/default.asp
Provides articles explaining tax topics, including many tips for reducing taxes.

Taxes: Tools and Reference
www.hrblock.com/taxes/tools/index.html
Includes daycare expenses, self-employment, withholding calculator, and more.

Online Discussions

misc.taxes.moderated
www.groups.google.com/groups?oi=djq&as_ugroup=misc.taxes
Discussion of taxes, proposed and existing tax laws, regulations, and procedures.

us.taxes (unmoderated)
www.groups.google.com/groups?oi=djq&as_ugroup=us.taxes
Discussion of U.S. taxes, policy, tax collections, and more.

Subject Headings

- finance, personal—United States
- income tax—law and legislation—United States
- income tax—United States
- tax planning—United States
- tax returns—United States—forms

U.S. Presidents: A Research Guide

Researching a specific U.S. president? Look no further. This guide lists some of the best books and Web sites for beginning researchers and information on how to find more. If you are looking for personal information, election results, cabinet members, and other areas of interest on each president, use the resources below to get started. The terms and phrases listed in the subject headings below can be used to search for more materials in the library's catalog and research databases. If you need further assistance, please ask a librarian.

Books

The American Heritage Illustrated History of the Presidents by Michael R. Beschloss. New York: Crown, 2000. ISBN: 0812932498.
Covers the history behind each presidency, through the end of the Clinton Administration. Provides photographs, political cartoons, and complete listings of Cabinet members and other administrative officials.

The Complete Book of U.S. Presidents, 5th ed. by William A. DeGregorio. New York: Gramercy Books, 2002. ISBN: 0517183536.
Covers the lives and accomplishments of the U.S. Presidents, through Clinton. Provides portraits and photographs of each.

Our Country's Presidents by Ann Bausum. Washington, DC: National Geographic Society, 2001. ISBN: 0792272269.
A full-color volume of presidential biography, up to George W. Bush, Jr.

Scholastic Encyclopedia of the Presidents and their Times by David Rubel. New York: Scholastic Reference, 1994. ISBN: 0590493663.
This illustrated encyclopedia gives brief information about each president, through Clinton. Second half of the book explores the social, cultural, and historical events of each president's time.

The Vice Presidents: A Biographical Dictionary edited by L. Edward Purcell. New York: Facts On File, 1998. ISBN: 0816031096.
A biographical dictionary of our country's vice presidents up to Albert A. Gore.

The White House: Cornerstone of a Nation by Judith St. George. New York: Putnam, 1990. ISBN: 0399221867.
Provides an overview of White House history, with time lines in the front. Illustrated.

The White House: An Historic Guide, 15th ed. White House Historical Association; National Geographic Society (United States). Washington, DC: The Association, 1982. ISBN: 0912308176.
An older book, this publication is nonetheless valuable as an inside look into the White House, its rooms, art, inhabitants, and development. Full-color illustrations.

Reference Books

American National Biography edited by John Arthur Garraty and Mark C. Carnes. 24 vols. New York: Oxford University Press, 1999. ISBN: 0195206355.
An alphabetical listing of notable people in American history, with biographical information.

Facts about the Presidents: A Compilation of Biographical and Historical Information, 6th ed. edited by Joseph Nathan Kane. New York: H. W. Wilson, 1993. ISBN: 0824208455.
A concise reference book that details the life, family, education, political histories and accomplishments of the presidents, up through George W. Bush, Jr. Also gives time lines and information on other top officials through presidential history.

Who's Who in America. 3 vols. Chicago: A. N. Marquis, 2004. ISSN: 00839396. Alphabetical listing of information on notable people in America throughout history. All the presidents will be in here.

Web Sites

American Presidents: Life Portraits
www.americanpresidents.org/gallery/
Presidential portraits from Washington to Bush by artist Chris Fagan.

Presidents of the United States (POTUS)
www.potus.com
A comprehensive Web site that offers many informative links. Scroll down and click on the president that you are researching. The bottom of the page offers a helpful, alphabetical subject index. Watergate, under "W," for instance, would provide a link to information on this subject, as well as a link to President Nixon's page.

Presidents of the United States (White House)
www.whitehouse.gov/history/presidents/
Brief biography and portraits from the book *The Presidents of the United States of America* written by Frank Freidel and Hugh S. Sidey (contributing author), published by the White House Historical Association with the cooperation of the National Geographic Society.

The White House
www.whitehouse.gov
This is the official Web site of the White House. For information on the presidents, click on "History and Tours," along the top of page. On the next page click on the "Presidents" link underneath the heading "Presidents and First Ladies." For information on the White House itself, click on "Life in the White House" under the heading "White House." On this page, you may view stories and pictures of rooms within the White House. These links are located on the right-hand side of the page.

Subject Headings

- presidents—United States—autographs
- presidents—United States—biography
- presidents—United States—correspondence
- presidents—United States—election (year)
- presidents—United States—history
- Add "juvenile literature" to retrieve materials for children.

U.S. Supreme Court: A Research Guide

Decisions from the Supreme Court affect our daily lives such as rulings from court cases about adoption, disability, discrimination, and employment. The resources listed below will help researchers gather historical information on the justices and the court, as well as provide current background information about cases currently being decided. The terms and phrases listed in the subject headings below can be used to search for more materials in the library's catalog and research databases. If you need further assistance, please ask a librarian.

Books

Encyclopedia of the American Constitution, 2nd ed. edited by Leonard Williams Levy, et al. New York: Macmillan Reference USA, 2000. ISBN: 0028648803.
Includes articles covering concepts and court cases since 1992 on topics like adoption, race, and the Constitution, birthright citizenship, Clinton v. Jones, disability discrimination, hate crimes, modern militias, Violence Against Women Act, and more.

Encyclopedia of the U.S. Supreme Court edited by T. T. Lewis and R. L. Wilson. 3 vols. Pasadena, CA: Salem Press, 2001. ISBN: 0893560979.
Essays on Supreme Court decisions; historical overviews of how the Court has treated important issues; specific historical events and eras; court administration and structure; legal terms; judicial interpretation of state and federal laws; and biographies of Supreme Court justices.

The Oxford Companion to the Supreme Court of the United States edited by Kermit Hall. New York: Oxford University Press, 1992. ISBN: 0195058356.
Includes more than a thousand alphabetically arranged entries including biographies of justices, rejected nominees, presidents who had an important impact on—or conflict with—the Court, and articles on major decisions.

The Supreme Court, A to Z, 2nd ed. edited by Kenneth Jost. Washington, DC: Congressional Quarterly, 1998. ISBN: 156802357X.
Alphabetical encyclopedia of over 300 essays includes information on significant decisions of the Supreme Court, the history of the Court, the justices, and the powers of the Court.

The Supreme Court of the United States: Its Beginnings and Its Justices, 1790–1991. Commission on the Bicentennial of the United States Constitution. Washington, DC: Commission on the Bicentennial of the United States Constitution, 1992.
Provides historical background on the chief justices, associate justices, homes of the Supreme court, a ceremony at the Supreme Court, members of the Supreme Court and the succession of the justices.

Web Sites

Federal Legal Information Through Electronics
www.fedworld.gov/supcourt/
Search and view full text of Supreme Court decisions issued between 1937 and 1975 Contains 7,407 Decisions from volumes 300 through 422 of U.S. Reports.

Landmark Supreme Court Cases
www.landmarkcases.org
Curriculum support for educators that includes background summaries, diagrams, excerpts from the majority and dissenting opinions, and full text of the Supreme Court's decisions.

Supreme Court Collection, Cornell University
http://supct.law.cornell.edu/supct/
Archives U.S. Supreme Court decisions dating back to 1990. Offers highlights on current term cases.

U.S. Supreme Court
www.supremecourtus.gov
Authoritative site offering information about the U.S. Supreme Court, docket of the court, oral arguments, court rules, and opinions.

U.S. Supreme Court Multimedia
www.oyez.org/oyez/frontpage/
Provides access to more than 2,000 hours of Supreme Court audio since 1995. Prior to that date a selective collection is available.

Organizations

Supreme Court Historical Society
www.supremecourthistory.org
The Supreme Court Historical Society is dedicated to the preservation and dissemination of the history of the Supreme Court of the United States.

Subject Headings

- Constitutional law—United States—encyclopedias
- judges—United States—biography
- United States Supreme Court
- United States Supreme Court—biography
- United States Supreme Court—encyclopedias
- United States Supreme Court—history

Weather: A Research Guide

Researchers and students search for weather information for a variety of reasons including creating travel plans and writing research reports. Weather affects farming, transportation, and business in many different ways since, at any given time, the weather can change. This guide introduces researchers to some of the basic informational sources on the topic. The resources below will help researchers get started if they are looking for forecasts for the next few days or even past weather conditions. The terms and phrases listed in the subject headings below can be used to search for more materials in the library's catalog and research databases. If you need further assistance, please ask a librarian.

Forecasts

Intellicast.com
www.intellicast.com
Includes recent historical data by locality.

Weather Channel
www.weather.com/homepage.html
Includes 10-day weather forecasts and how weather impacts travel, lawn and garden, health, health, and recreation.

Weather Underground
www.wunderground.com
Provides daily weather forecast and includes historical data by locality back to the 1920s in some cases, such as Boston.

Reference Books

Encyclopedia of Climate and Weather by Stephen Henry Schneider. New York: Oxford University Press, 1996. ISBN: 0195094859.
With more than 300 entries covering topics from acid rain to zonal circulation though text and more than 400 illustrations.

The Encyclopedia of Climatology edited by John E. Oliver and Rhodes Whitmore Fairbridge. New York: Van Nostrand Reinhold, 1987. ISBN: 0879330090.
Supplies data on climates in major continental areas. Explains causes of climatic processes and changes in more than 200 articles.

Macmillan Encyclopedia of Weather by Paul Stein. New York: Macmillan Reference USA, 2001. ISBN: 0028654730.
Includes 150 entries covering weather in all its manifestations, terms, concepts, and particular phenomena in a single volume. Introductory work for students and general researchers.

The Weather Almanac: A Reference Guide to Weather, Climate, and Air Quality in the United States and Its Key Cities, Comprising Statistics, Principles, and Terminology. Detroit: Gale Research, 2003. ISBN: 0787675156.
Includes in-depth weather records for 108 major U.S. cities and a climatic overview of the country, including 33 U.S. weather atlas maps.

Weather America by Alfred N. Garwood. Milpitas, CA: Toucan Valley, 1996. ISBN: 188492560X.
Includes detailed climatological data for over 4,000 places and rankings.

Weather America: A Thirty-Year Summary of Statistical Weather Data and Rankings, 2nd ed. by David Garoogian. Lakeville, CT: Grey House, 2001. ISBN: 1891482297.
Covers weather rankings, major storm events, and state chapters. Appendices include climate centers, periods of record, map of NEXRAD Doppler radar stations, and an explanation of data.

Organizations

National Climatic Data Center
http://lwf.ncdc.noaa.gov/oa/ncdc.html
Produces numerous climate publications and responds to data requests from all over the world. NCDC supports a three tier national climate services support program—the partners include: NCDC, Regional Climate Centers, and State Climatologists. Web site includes many free data files.

National Weather Association (NWA)
www.nwas.org
Promotes professionalism and develops solutions to problems faced by meteorologists and people working in daily weather forecasting activities.

National Weather Service, National Oceanic and Atmospheric Administration
www.nws.noaa.gov
Issues weather warnings, observations, forecasts, hydrological information reported from 13 forecast centers, and satellite maps.

Subject Headings

- geographic—United States—climate—tables
- meteorology—dictionaries
- meteorology—United States—tables
- weather—dictionaries

Winter Olympics: A Research Guide

The materials listed below will entertain and inform you about the Winter Olympics. With a history that can be traced back to ancient Greece, this topic is both interesting and exciting. Since 1994, the winter games have been held in even-numbered years in which the summer games are not contested. Olympian athletes dream of winning medals while representing their country and viewers cheer them on in excitement. Some of the winter sports include ice hockey, curling, bobsledding, luge, skeleton, skiing, snowboarding, and skating events. Additional information can be found on Olympians by searching their name in the library catalog or your favorite search engine. You can also search for magazine and newspaper articles. If you need further assistance, please ask a librarian.

Books

The Encyclopedia of the Winter Olympics by John F. Wukovits. New York: F. Watts, 2001. ISBN: 0531118851.
Covers the history, origin of each sport, equipment design, and its bearing on performance.

The Olympic Winter Games by Caroline Arnold. New York: F. Watts, 1991. ISBN: 0531200531.
A brief history of the winter Olympic games, with descriptions of individual events and profiles of several past champions.

The Winter Olympics by Larry Dane Brimner. New York: Children's Press, 1997. ISBN: 0516204564.
Beginning with the Nordic Games in 1908, this book covers some of the sports involved, including skiing, ice hockey, skating, and bobsledding.

The Winter Olympics by Jack C. Harris. Mankato, MN: Creative Education, 1990. ISBN: 088682317X.
Discusses the history of the Winter Olympics, the events, and outstanding athletes over the years.

Web Sites

International Olympic Committee
www.olympic.org
Official site of the International Olympic Committee (IOC).

Salt Lake City 2002
www.saltlake2002.com
Official site of the Salt Lake City 2002 Winter Olympics. Site includes event and athlete information.

Torino, Italy 2006
www.torino2006.it/eng/index.asp
Official site of the XX Olympic Games in Torino, Italy 2006.
Information about the torch route through Indianapolis including a map of the route and information about local torchbearers.

United States Olympic Committee
www.olympic-usa.org
United States Committee that organizes and promotes the games for team USA.

United States Olympic Team
www.usolympicteam.com
Biographies of current and past athletes of the Olympics.

OLYMPIC HISTORY

Coubertin Museum
www.museum.olympic.org/e/gallery/permanent/cou_bio_e.html
Museum that highlights the life and work of Pierre Frédy, Baron de Coubertin, the father of the modern Olympics.

Coubertin's Olympics: How the Games Began by Davida Kristy. Minneapolis: Lerner, 1995. ISBN: 0822533278.
Traces the history of modern day Olympics back to Pierre de Coubertin (1863–1937).

History of Olympic Mascots
www.collectors.olympic.org/e/fimo/fimo_mascots_e.html
Everything you ever wanted to know about Olympic mascots beginning with "Schuss," the first unofficial Olympic Mascot in Grenoble in 1968 and "Waldi the Dachshund," the first official mascot in Munich in 1972 to the present day.

The Nazi Olympics: Berlin 1936 by Susan D. Bachrach. Boston: Little, Brown, 2000. ISBN: 0316070866.
Recounts the story of the Olympics held in Berlin in 1936.

The Olympic Festival in Antiquity
http://sunsite.icm.edu.pl/olympics/classical/intro.html
Learn about the people, places, rules, and games of the Ancient Olympics.

Olympic Museum
www.museum.olympic.org
Learn about Coubertin and Olympic History, explore a library of Olympic images and sounds, and museum exhibits.

Perseus Project: The Olympic Games
www.perseus.tufts.edu/Olympics/
Compare ancient and modern Olympic sports, tour the site of Olympia as it looks today, and learn about the ancient Olympics.

The Story of the Olympics by Dave Anderson. New York: W. Morrow, 1996. ISBN: 0688129544.
History of the Olympics dating back to 776 B.C. Includes stories of particular events such as track and field, gymnastics, and speed skating.

SUBJECT HEADINGS

- athletes—biography—juvenile literature
- Olympics—encyclopedias
- Olympics—history
- Olympics—history—juvenile literature
- Olympics—juvenile literature
- Winter Olympics—encyclopedias
- Winter Olympics—encyclopedias—juvenile literature
- Winter Olympics—history
- Winter Olympics—history—juvenile literature

Women's Studies: A Research Guide

Whether studying current issues in women's studies or historical biographies and how women from the past impact our lives today, researchers will find a great deal of information in the materials listed below. Students and researchers can start their research here and use terms and phrases listed in the subject headings below to search for more materials in the library's catalog and research databases. If you need further assistance, please ask a librarian.

Biographical Sources

Notable American Women, 1607–1950: A Biographical Dictionary edited by Edward T. James, Janet Wilson James, and Paul S. Boyer. Cambridge, MA: Belknap Press of Harvard University Press, 1971. ISBN: 0674627318.
Provides biographies of notable women from abolitionists to welfare work leaders, wives of the presidents, and women's club leaders Prepared under the auspices of Radcliffe College.

Notable Black American Women by Jessie Carney Smith and Shirelle Phelps. 3 vols. Detroit: Gale Research, 1992–2003. ISBN for vol. 1: 0810347490; ISBN for vol. 2: 0810391775; ISBN for vol. 3: 0787664944.
Narrative biographical essays discuss each woman's significant achievements. Includes over 700 entries.

Notable Hispanic American Women edited by Diane Telgen and Jim Kamp. Detroit: Gale Research, 1993. ISBN: 0810375788; 1998, ISBN: 0787620688.
Provides informative profiles of more than 500 Hispanic-American women and includes personal, career, and educational information and discuss significant achievements.

Who's Who of American Women. Chicago: Marquis Who's Who, 1971– . Biennial. ISSN: 00839841. (1959–1965, ISSN: 02702940).
Over 33,000 entries of women professional credentials of women chosen because of their professional accomplishments, including local and federal elected and appointed officials, educators, recipients of awards, and more.

Women in World History: A Biographical Encyclopedia by Anne Commire and Deborah Klezmer. 17 vols. Waterford, CT: York, 1999–2002. ISBN: 078763736X.
More than 2,500 signed articles written by academics, up to 5,000 words in length of women from many nations both living and dead. Articles vary in length according to the subject because in some cases only birth, marriage, children, and death dates are known.

Encyclopedias

Encyclopedia of Women's History in America, 2nd ed. by Kathryn Cullen-DuPont. New York: Facts On File, 2000. ISBN: 0816041008.
Includes 500 entries that highlight prominent women and how they contributed to society throughout history.

Handbook of American Women's History by Angela Howard and Frances M. Kavenik. New York: Garland, 1990. ISBN: 0824087445.
Articles on important people, events, and ideas that have shaped the history of women in the United States.

Women's Issues edited by Robin Brown. New York: H. W. Wilson, 1993. ISBN: 0824208447.
Covers topics like women in the workplace, sexual harassment, work and family, and politics.

Women's Studies Encyclopedia edited by Helen Tierney. 3 vols. New York: Greenwood Press, 1989–1991. ISBN: 0313246467.
Signed articles from each volume include: vol. 1. Views from the Sciences; vol. 2. Literature, Arts, and Learning; vol. 3. History, Philosophy, and Religion.

Statistical Resources

Statistical Abstract of the United States. Washington, DC: Government Printing Office, 1878– . ISSN: 00814741. Available on the Internet at www.census.gov/statab/www/. Online 1995–present.
As the National Data Book, it contains a collection of statistics from abortions, age, death, and birth rate to volunteers, voter registration, and turnout.

Statistical Handbook on Women in America, 2nd ed. edited by Cynthia Murray Taeuber. Phoenix: Oryx Press, ISBN: 1573560057.
Statistical record of women worldwide.

Subject Headings

- African American women—biography
- blacks—biography
- blacks—United States—biography
- feminism—encyclopedias
- Hispanic-American women—biography—dictionaries
- women—biography
- women—encyclopedias
- women—United States—biography—dictionaries
- women—United States—encyclopedias
- women political activists—encyclopedias
- women social reformers—encyclopedias
- women's rights—encyclopedias

World History: A Research Guide

History scholars piece together the past by uncovering information through a variety of sources, some obscure and others standard. This guide introduces researchers to some of the basic informational sources on the topic. The terms and phrases listed in the subject headings below can be used to search for more materials in the library's catalog and research databases. If you need further assistance, please ask a librarian.

Books

The American Historical Association's Guide to Historical Literature, 3rd ed. edited by Mary Beth Norton and Pamela Gerardi. 2 vols. New York: Oxford University Press, 1995. ISBN: 0195057279.

Guide to historical literature for all periods and areas of the world. Especially strong on European history, with sections on individual countries, the expansion of Europe, World Wars I and II, and international relations.

Chronology of World History: A Calendar of Principal Events from 3000 B.C. to A.D. 1976, 2nd ed. edited by Greville Stewart Parker Freeman-Grenville. London: Rex Collins, 1978. ISBN: 0901720674.

Chronology of world events including coverage of non-European history.

Dictionary of Concepts in History edited by Harry Ritter. Westport, CT: Greenwood Press, 1986. ISBN: 0313227004.

Includes key concepts of contemporary historical analysis. Entries include a brief definition of the concept and a summary of its history, a list of references cited, and a bibliography of sources for further reading.

A Dictionary of Nineteenth-Century World History edited by John Belchem and Richard Price. Cambridge, MA: Blackwell, 1994. ISBN: 0631183523.

Includes 800 entries covering all aspects of political, diplomatic, military, social, and economic history, and provides overviews of the cultural and artistic developments of the period.

The Encyclopedia of World History: Ancient, Medieval, and Modern, Chronologically Arranged, 6th ed. edited by Peter N. Stearns and William L. Langer. Boston: Houghton Mifflin, 2001. ISBN: 0395652375.

Chronology of more than 20,000 entries that span the millennia from prehistoric times to the year 2000 grouped by geographic regions. Completely revised in this expanded edition.

Everyman's Dictionary of Dates, 7th ed. (revised) edited by Audrey Butler. London: J. M. Dent & Sons, 1987. ISBN: 0460029053.

Chronology of world events in dictionary format with entries under the names of countries.

Historical Journals: A Handbook for Writers and Reviewers, 2nd ed. edited by Dale R. Steiner and Casey R. Phillips. Santa Barbara, CA: ABC-CLIO, 1993. ISBN: 0899508014.

A guide for writers interested in placing works for publication in historical journals. Brief sections offering advice on articles and book reviewing are followed by an alphabetical listing of about 700 historical journals.

Historical Periodicals Directory edited by Eric H. Boehm, Barbara H. Pope, and Marie Ensign. 5 vols. Santa Barbara, CA: ABC-CLIO, 1981–1986. ISBN: 0874360188.

Guide to historical periodicals.

The Modern Researcher, 6th ed. edited by Jacques Barzun and Henry F. Graff. Belmont, CA: Thomson/Wadsworth, 1992. ISBN: 0155055291.
Introduction to the techniques of research and the art of expression is used widely in historiography.

Reference Sources in History: An Introductory Guide, 2nd ed. by Ronald H. Fritze, Brian E. Coutts, and Louis Andrew Vyhnanek. Santa Barbara, CA: ABC-CLIO, 2004. ISBN: 0874368839.
An annotated bibliography of reference materials useful to historians, including bibliographies, periodical guides and indexes, newspapers, statistical sources, government publications, and biographical sources.

Sources of Information for Historical Research edited by Thomas P. Slavens. New York: Neal-Schuman, 1994. ISBN: 1555700934.
Sources organized by geography (country or continent), genre, and type of holding. All works listed are annotated. The imprints are primarily in English and are arranged by LC classification number. Access is further enhanced by author, title, and subject indexes.

Women in World History: A Biographical Encyclopedia. 15 vols. Anne Commire and Deborah Klezmer. Detroit: Gale Group, 1999– . ISBN: 07863736X.
Over 10,000 articles about individual women from many nations both living and dead.

Journals

American Historical Review
China Quarterly
History in Africa
International Journal of Middle East Studies
Journal of Medieval History
Middle Eastern Studies
Modern Asian Studies
Russian Studies in History
Slavic Review
William and Mary Quarterly
Yearbook of German-American Studies

Web Sites

HistoryWorld
www.historyworld.net
400 separate historical articles and descriptions of approximately 4,000 world events. Emphasis on English history.

WebChronology Project
http://campus.northpark.edu/history/WebChron/
A multidimensional chronology. Uses Web links to illustrate the relationship between long term and short term events and view the relation between events in one field to events in another.

Women in World History
www.womeninworldhistory.com
Full of information and resources to help you learn about women's history in a global, non-U.S., context.

Subject Headings

Tips for constructing subject searches: Include keywords that indicate the time period, geographic period and the type of information sought (i.e., encyclopedias, dictionaries).

- civilization—history—encyclopedias
- history—outlines, syllabi, etc.
- history, modern—19th century
- history, modern—20th century
- military history, modern—20th century
- revolutions—history—20th century

World War I: A Research Guide

Researchers of the Great War will find a good deal of information to jump start their research from the resources listed below. From the assassination of Archduke Franz Ferdinand and the events preceding it to the signing of the Versailles Treaty in 1919. The information is presented in a variety of formats including personal narratives, chronologies, primary documents, photographs, and encyclopedias. The terms and phrases listed in the subject headings below can be used to search for more materials in the library's catalog and research databases. If you need further assistance, please ask a librarian.

Recommended Books

Almanac of World War I by David F. Burg and L. Edward Purcell. Lexington, KY: University Press of Kentucky, 1998. ISBN: 0813120721.
A daily account of the action on all fronts including daily entries, topical descriptions, biographical sketches, maps, and illustrations.

Chronicle of the First World War by Randal Gray and Christopher Argyle. 2 vols. New York: Facts On File, 1991. ISBN for vol. 1, 1914–1916: 0816021392; ISBN for vol. 2, 1917–1921: 0816025959.
Chronicles military, political, and international aspects of WWI.

The First World War by John Keegan. New York: Knopf, 1999. ISBN: 0375400524.
Definitive narrative account of World War I by notable historian.

The First World War: A Complete History by Martin Gilbert. New York: Henry Holt, 1994. ISBN: 080501540X.
Acclaimed British author covers the entire length of the war in this readable one volume book detailing the seeds of war and how it changed the world.

The First World War: An Eyewitness History by Joe H. Kirchberger. New York: Facts On File, 1992. ISBN: 0816025525.
Provides first-hand accounts of the war in memoirs, speeches, letters, and newspapers. Appendixes include 56 documents, more than 200 brief biographies, and five maps, bibliography, index, and 75 photographs.

Historical Atlas of World War I by Anthony Livesey and H. P. Willmott New York: Henry Holt, 1994. ISBN: 0805026517.
Primarily of battle maps in color with commentary and time lines arranged in chronological order.

History of World War I. 3 vols. New York: Marshall Cavendish, 2002. ISBN: 0761472312.
Articles are arranged in three volumes, providing a chronological account and regional overviews of the worldwide conflict. "War and Response, 1914–1916" (vol. 1), "Victory and Defeat, 1917–1918" (vol. 2), and "Home Fronts, Technologies of War" (vol. 3).

My Experiences in the World War by John J. Pershing. 2 vols. New York: Frederick A. Stokes Company, 1931.
First-hand account of author's time spent as Commander-in-Chief of the American Expeditionary Forces in World War I.

Web Sites

Encyclopaedia of the First World War
www.spartacus.schoolnet.co.uk/FWW.htm
Provides comprehensive coverage of the war in an easy to navigate format.

The Great War and the Shaping of America
www.pbs.org/greatwar/index.html
Includes a great deal of information, as well as interactive features.

The World War I Document Archive
www.lib.byu.edu/estu/wwi/
Archive of primary documents from World War I. Includes image archive.

KEYWORDS

Try these search terms in the library's online databases. Use print indexes for articles published before online databases coverage began.

Balkan Wars, 1912–1913
Fourteen Points
League of Nations
Lusitania, Sinking of the, 1915
Russia—Revolution, 1917–1921
Triple Alliance or Triple Entente
Verdun, Battle of
Versailles, Treaty of, 1919
William II, Emperor of Germany
Wilson, Woodrow
World War, 1914–1918

SUBJECT HEADINGS

- World War, 1914–1918
- World War, 1914–1918—campaigns
- World War, 1914–1918—children's fiction
- World War, 1914–1918—fiction
- World War, 1914–1918—participation, African-American
- World War, 1914–1918—United States
- For biographies, search under the person's name, e.g., Hitler, Adolf or Patton, George.

World War II: A Research Guide

Students of World War II will find a great deal of information to begin their research from the resources listed below. Resources include a wide array of coverage from broad topics like the origin of war to specific ones like the Munich Pact. The information is presented in a variety of formats including personal narratives, chronologies, primary documents, photographs, oral histories, and general encyclopedias. The terms and phrases listed in the subject headings below can be used to search for more materials in the library's catalog and research databases. If you need further assistance, please ask a librarian.

Books

Biographical Dictionary of World War II by Mark Mayo Boatner. Novato, CA: Presidio Press, 1996. ISBN: 0891415483.
Authoritative accounts of more than 1,000 key personalities from the war years.

A Dictionary of the Second World War by Elizabeth-Anne Wheal, Stephen Pope, and James Taylor. New York: P. Bedrick Books, 1990. ISBN: 0872263371.
Covers more than 1,600 detailed and analytical entries including theaters, weapons, tactics and strategies and politicos and diplomacy.

Encyclopedia of the Holocaust by Israel Gutman. 2 vols. New York: Macmillan Reference USA, 1995. ISBN: 0028645278.
Covers all aspects of the Holocaust in over 1,000 articles. Includes photographs, drawings, and maps.

The Greatest Generation by Tom Brokaw. New York: Random House, 1998. ISBN: 0375502025.
Personal narratives of Americans born in the 1920s who came of age during the Great Depression and fought in World War II.

Louis L. Snyder's Historical Guide to World War II. Westport, CT: Greenwood Press, 1982. ISBN: 0313232164.
Brief entries cover individuals, concepts, and events of the economic, social, cultural, psychological, political, and military aspects of the war.

The Oxford Companion to World War II by Ian Dear and M. R. D. Foot. New York: Oxford University Press, 1995. ISBN: 0198662254.
Covers all aspect of the conduct and experience of the war in over 1,700 entries ranging from brief identifications to in-depth articles on complex subjects from 140 experts.

Atlases

Historical Atlas of the Holocaust. United States Holocaust Memorial Museum. New York: Macmillan, 1996. ISBN: 0028974514.
This book features maps that illustrate Europe before the war; the Holocaust in Eastern Europe; Western, Central Europe, Southern Europe, and Hungary; Nazi extermination camps; Jewish armed resistance; death marches; liberation; and postwar Europe from 1945–1950.

The Historical Atlas of World War II by John Pimlott and Alan Bullock. New York: Henry Holt, 1995. ISBN: 0805039295.
This atlas chronicles the major and minor campaigns of the war in Europe, Africa, the Middle East, Asia, and the Pacific in over one hundred full-color maps with more than one hundred captioned color and black-and-white photographs.

The Times Atlas of the Second World War by John Keegan. New York: Harper & Row, 1989. ISBN: 0060161787.
This atlas plots the exact course, on ground, at sea, and in the air, of the Second World War through hundreds of maps, charts, and commentaries.

Web Sites

Experiencing War: Stories from the Veteran's History Project
www.loc.gov/folklife/vets/stories/ex-war-home.html
Personal narratives recorded from veterans of various wars.

The Perilous Fight: America's World War II in Color
www.pbs.org/perilousfight/
Combines original color film footage with compelling passages from diaries and letters.

Second World War Encyclopaedia
www.spartacus.schoolnet.co.uk/2WW.htm
Includes a great deal of general information but also information not readily available, such as war photographers, journalists, and artists.

A Teacher's Guide to the Holocaust
http://fcit.coedu.usf.edu/holocaust/
An overview of the people and events of the Holocaust through photographs, documents, art, music, movies, and literature.

World War II Historical Text Archive
www.historicaltextarchive.com
Choose from full text articles, e-books, and links about World War II.

World War II: The Homefront
http://library.thinkquest.org/15511/
A great resource including a time line in photographs, an online museum displays medals and memorabilia, and much more. Great resource for self-guided study.

World War II Timeline
http://history.acusd.edu/gen/WW2Timeline/start.html
Chronology of events in words and photos.

Subject Headings

- Americans—evacuation and relocation, 1942–1945
- Holocaust—Jewish (1939–1945)
- Japanese-Americans—evacuation and relocation, 1942–1945
- World War, 1939–1945
- World War, 1939–1945—aerial operations (naval operations)
- World War, 1939–1945—biography
- World War, 1939–1945—fiction
- World War, 1939–1945—Jews
- For biographies, search under the person's name, e.g., Hitler, Adolf or Patton, George.

Bibliography

Annis, Eleanor. 1997. "Pathfinders—Revisited." *Arkansas Libraries* 54: 9–12.

Barnes, Jeanne. 2003. "A Pathfinder for Constructing Pathfinders." Wenatchee School District Home Page. (April 30) Available: http://home.wsd.wednet.edu/pathfinders/path.htm.

Borne, Barbara Wood. 1996. *100 Research Topic Guides for Students*. Westport, CT: Greenwood.

Canfield, Marie P. 1972. "Instructional Materials: Design and Development." Library Pathfinders. *Drexel Library Quarterly* 8, no. 3 (July): 287–300.

Covert, Kay. 2001. "How the OCLC CORC Service is Helping Weave Libraries into the Web." *Online Information Review* 25, no. 1: 41–46.

Dahl, Candice. 2001. "Electronic Pathfinders in Academic Libraries: An Analysis of Their Content and Form." *College and Research Libraries* 62, no. 3 (May): 227–237.

Dunsmore, Carla. 2002. "A Qualitative Study of Web-Mounted Pathfinders Created by Academic Business Libraries." *Libri* 52, no. 3: 137–156.

Gangl, Susan. 2001. "The Librarian's Library: Fugitive Reference Files." *Reference Librarian* 72: 179–195.

Graves, Judith K. 1998. "Research Pathfinders: Offline Access to Online Searching." *Multimedia Schools* 5, no. 3 (May/June): 26–30.

Grealy, Deborah S. 2000. "Technological Mediation: Reference and the Non-Traditional Student." *Reference Librarian* 69/70: 63–69.

Grimes, Marybeth, and Sara E. Morris. 2001. "A Comparison of Academic Libraries' Webliographies." *Internet Reference Services Quarterly* 5, no. 4: 69–77.

Harbeson, Eloise L. 1972. "Teaching Reference and Bibliography: The Pathfinder Approach." *Journal of Education for Librarianship* 13, no. 2: 111–115.

Hernon, Peter. 1974. "Pathfinders/Bibliographic Essays and the Teaching of Subject Bibliography." *RQ* 13, no. 3: 235–238.

Holtze, Terri L., and Anna Marie Johnson. 1997. "Getting Mileage out of the Pathfinder." *Kentucky Libraries* 61 (Spring): 29–32.

Jacso, Peter. 2003. "Create Digitally Enhanced Bibliographies with Public Domain Databases." *Computers in Libraries* 23 (June): 52–54.

Jarvis, William E. 1985. "Integrating Subject Pathfinders into Online Catalogs." *Database* 8 (February): 65–67.

Jarvis, William E., and Victoria E. Dow. 1986. "Integrating Subject Pathfinders into a GEAC ILS (Integrated Library System): A MARC-Formatted Record Approach." *Information Technology and Libraries* 5 (September): 213–227.

Jung, Claudia Ruediger, et al. *Pathfinders: An Intellectual Guide to Libraries.* Castleton, VT: Castleton State College. ED287484.

Kapoun, Jim M. 1995. "Re-Thinking the Library Pathfinder." *College and Undergraduate Libraries* 2, no. 1: 93–105.

Katz, Janet C. 2001. "No One Person: Views on a Collaboration." *Legal Reference Services Quarterly* 20, no. 3: 105–121.

Kuntz, Kelly. 2003. "Pathfinders: Helping Students Find Paths to Information." *Multimedia Schools* 10, no. 3 (May/Jun): 12–16.

Laverty, Corinne Y. C. 1997. "Library Instruction on the Web: Inventing Options and Opportunities." *Internet Reference Services Quarterly* 2, nos. 2–3: 55–66.

McDougald, Dana. 1999. *100 More Research Topic Guides for Students.* Westport, CT: Greenwood.

Miller, Donna. 2000. "Library Pathfinders." *Library Talk* 13, no. 2 (March/April): 20–22.

Morris, Sara E., and Marybeth Grimes. 1999. "A Great Deal of Time and Effort: An Overview of Creating and Maintaining Internet-Based Subject Guides." *Library Computing: Internet and Software Applications for Information Professionals* 18, no. 3 (September): 213–217.

Nielsen, Jakob. "Usability 101" (August 25, 2003). Accessed on January 28, 2004. Available: www.useit.com/alertbox/20030825.html.

Nuttall, Harry D., and Sonja L. McAbee. 1997. "Pathfinders On-Line: Adding Pathfinders to a NOTIS On-Line System." *College and Undergraduate Libraries* 4, no. 1: 77–101.

O'Sullivan, Michael K., and Thomas J. Scott. 2000. "Pathfinders Go Online." *School Library Journal* (Summer): S40–42.

Peterson, Lorna, and Jamie Wright Coniglio. 1987. "Readability of Selected Academic Library Guides." *RQ* 27 (Winter): 233–239.

Poulter, Alan. 1987. "Library Guides: Printed and Electronic." *Audiovisual Librarian* 13 (May): 110.

Reference and User Services Association (RUSA), ALA. 2001. *Guidelines for the Preparation of a Bibliography.* Available: www.ala.org.

Rogers, Robert. 1985. "A Comparison of Manual and Online Searches in the Preparation of Philosophy Pathfinders." *Journal of Education for Library and Information Science* 26 (Summer): 54–55.

Shearer, Barbara, et al. "Bibliographic Instruction through the Related Studies Division in Vocational Education: LRC Guide, Pathfinders, and Script for Slide Presentation." Tennessee Area Vocational Technical Schools. ERIC. ED205171.

Stevens, Charles H., et al. 1973. "Library Pathfinders: A New Possibility for Cooperative Reference Service." *College and Research Libraries* 34, no. 1 (January): 40–46.

Subject Bibliographies and Research Guides. 1985–1987. San Francisco State University: J. Paul Leonard Library. ERIC. ED296730.

Sutter, Lynne, and Herman Sutter. 1999. *Finding the Right Path: Researching Your Way to Discovery.* Worthington, OH: Linworth.

Taylor, Mary K., and Diane Hudson. 2000. "'Linkrot' and the Usefulness of Web Site Bibliographies." *Reference and User Services Quarterly* 39, no. 3 (Spring): 273.

Thompson, Glenn J., and Barbara R. Stevens. 1985. "Library Science Students Develop Pathfinders." *College and Research Libraries News* 46, no. 5 (May): 224–225.

Turner, Diane. 1993. "What's the Point of Bibliographic Instruction, Point-of-Use Guides and In-House Bibliographies?" *Wilson Library Bulletin* 67, no. 5 (January): 64–68.

Warner, Alice Sizer. "Pathfinders." *American Libraries* 14, no. 3 (March 1983): 150–152.

Wilbert, Shirley. 1981. "Library Pathfinders Come Alive." *Journal of Education for Librarianship* 21, no. 4 (Spring): 345–349.

Wilson, Paula. 2002. "Perfecting Pathfinders for the Web." *Public Libraries* 41, no. 2 (March/April): 99–100.

Wise, Maxine. 1990. "The Use of Modified Pathfinders to Facilitate Bibliographic Instruction in a Middle School." MSLS diss., University of North Carolina at Chapel Hill.

Yucht, Alice. 2001. "Project Pathfinder Portal Pages." *Library Talk* 14, no. 2 (March/April): 15–17.

Suggested Compilations of Pathfinders

School Libraries

1. African-American History
2. American History
3. Ancient Cultures
4. Ancient Egypt
5. Animals
6. Asian-Pacific American Heritage
7. Biographies
8. Biomes
9. Celebrations Around the World
10. Contemporary Issues
11. Countries
12. Current Events
13. Decades
14. Elections for Kids
15. Exploration and Discovery
16. Financial Aid and Scholarships
17. Hispanic Heritage
18. Inventors and Inventions
19. Literary Criticism
20. Mythology
21. Native American Heritage
22. Natural Disasters
23. On This Day in History
24. Research Tools
25. Saints
26. Science Experiments
27. Space Exploration
28. Space Shuttle
29. Summer Olympics
30. U.S. Civil Rights
31. U.S. Civil War
32. U.S. Congress
33. U.S. Constitution
34. U.S. Presidents
35. U.S. Supreme Court
36. Winter Olympics

Academic Libraries

1. Advertising
2. American Folklife
3. American History
4. Anthropology
5. Biographies
6. Biology
7. Book Reviews
8. Chemistry and Chemical Engineering
9. Company Information

Academic Libraries *(continued)*

10. Computer Science and Software Engineering
11. Criminal Justice
12. Demographics
13. Economics
14. Education
15. Engineering
16. English Literature
17. Environmental Studies
18. Gay and Lesbian Studies
19. Geography
20. Geology
21. Government Information
22. Linguistics
23. Literary Criticism
24. Military Science
25. Multicultural Literature
26. Music
27. Psychology
28. Public Administration and Policy
29. Quotations
30. Religion
31. Social Welfare
32. Theater
33. Women's Studies
34. World History
35. World War I
36. World War II

Public Libraries

1. Adoption
2. African-American History
3. Antiques and Collectibles
4. Asian-Pacific American Heritage
5. Automobile Repair
6. Biographies
7. Business Plans
8. Buying a Car
9. Careers and Jobs
10. Celebrations Around the World
11. College Search
12. Company Information
13. Consumer Information
14. Contemporary Issues
15. Cover Letters Resumes
16. Current Events
17. Decorative Arts Identification
18. Demographics
19. Diseases
20. Elections
21. Elections for Kids
22. English Literature
23. Etiquette
24. Financial Aid and Scholarships
25. Finding Forms
26. Finding People
27. Gardening
28. Genealogy
29. Government Information
30. Grants
31. Health and Wellness
32. Hispanic Heritage
33. Literary Criticism
34. Moving and Relocation
35. Mythology
36. Native American Heritage
37. Natural Disasters
38. Obituaries
39. On This Day in History
40. Patents, Trademark, and Copyright
41. Personal Finance
42. Quotations
43. Research Tools
44. Small Business Start-Up
45. Standards and Specifications

Public Libraries *(continued)*

46. Terrorism
47. Test Preparation
48. Travel Information
49. U.S. Income Tax Preparation
50. Weather

General Interest

1. Advertising
2. African-American History
3. American History
4. Animals
5. Asian-Pacific American Heritage
6. Biographies
7. Book Reviews
8. Career Information and Job Hunting
9. Celebrations Around the World
10. Company Information
11. Consumer Information
12. Contemporary Issues
13. Countries
14. Cover Letters and Resumes
15. Current Events
16. Decades
17. Demographics
18. Diseases
19. Elections
20. Elections for Kids
21. Etiquette
22. Finding Forms
23. Finding People
24. Gay and Lesbian Studies
25. Genealogy
26. Government Information
27. Grants
28. Health and Wellness
29 Hispanic Heritage
30. Legal Information
31. Moving and Relocation
32. Music
33. Native American Heritage
34. Obituaries
35. On This Day in History
36. Patents, Trademark, and Copyright
37. Personal Finance
38. Quotations
39. Religion
40. Research Tools
41. Saints
42. Travel Information
43. U.S. Congress
44. U.S. Constitution
45. U.S. Income Tax
46. U.S. Presidents
47. U.S. Supreme Court
48. Weather
49. Women's Studies
50. World History

Index

T

U

W

X

About the Author

Paula Wilson received an MLIS from the University of Rhode Island in 1992 and is currently the Web and Outreach Services Coordinator for the Maricopa County Library District (Maricopa, AZ). Previously, she worked at the Las Vegas-Clark County Library District, serving as Virtual Library Administrator, Reference Department head, and Reference Librarian. She began her library career at Providence Public Library (Providence, RI).

A contributing editor of the "Tech Talk" column in *Public Libraries*, Paula Wilson also authored *Library Web Sites: Creating Online Collections and Services* (ALA Editions, 2004). She maintains a Web site of her library-related work at *www.webliography.org.*